FLORIDA SCRUB-JAY

UNIVERSITY PRESS OF FLORIDA

Florida A&M University, Tallahassee
Florida Atlantic University, Boca Raton
Florida Gulf Coast University, Ft. Myers
Florida International University, Miami
Florida State University, Tallahassee
New College of Florida, Sarasota
University of Central Florida, Orlando
University of Florida, Gainesville
University of North Florida, Jacksonville
University of South Florida, Tampa
University of West Florida, Pensacola

Florida Scrub-Jay

Field Notes on a Vanishing Bird

MARK JEROME WALTERS

University Press of Florida

Gainesville · Tallahassee · Tampa · Boca Raton

Pensacola · Orlando · Miami · Jacksonville · Ft. Myers · Sarasota

Map drawings are courtesy of Derek Sarty.

ISBN 978-0-8130-6672-1
Library of Congress Control Number: 2020938187

The University Press of Florida is the scholarly publishing agency for the State University System of Florida, comprising Florida A&M University, Florida Atlantic University, Florida Gulf Coast University, Florida International University, Florida State University, New College of Florida, University of Central Florida, University of Florida, University of North Florida, University of South Florida, and University of West Florida.

University Press of Florida
2046 NE Waldo Road
Suite 2100
Gainesville, FL 32609
http://upress.ufl.edu

To my daughter, Anna, who sees
To my son, William, who believes
To my wife, Noelle, who knows

CONTENTS

ACKNOWLEDGMENTS

You can measure a book by pages and the time it took to write by years, but there is no measure for the company of those you love. Noelle, Will, and Anna, thank you for always being there. To all who made this work possible in other ways, I'm also indebted. Thanks to the staff and researchers at Archbold Biological Station, especially Reed Bowman, Nancy Chen, John W. Fitzpatrick, Mark A. Deyrup, Eric S. Menges, Vivienne Sclater, and Aurélie Coulon. You and many others elsewhere contributed your precious time with me in the field, in your offices, on the phone, or through correspondence, sharing your insights and expertise on the scrub-jay and its habitat: Dave Breininger, Steve Christman, Santiago Claramunt, Tony Clements, the late Sandy Cooper, Joel L. Cracraft, Xavier De Seguin Des Hons, Craig Faulhaber, Jay Garcia, David Gordon, Vivienne Handy, Albert Hine, Mary Keith, Mike Legare, Todd Mecklenborg, Karl Miller, Reed Noss, Brendan O'Connor, Michael J. Papa, Bruce Peterjohn, Katherine Prophet, Paul Schmalzer, Spencer Simon, and Jon Thaxton. More than sharing just your knowledge, you frequently shared your hopes and fears about the plight of the scrub-jay. This helped to make the book more than about an endangered bird. You spoke to the endangerment of human experience. My gratitude to Derek Sarty for contributing the beautiful section-maps under the Pal's Act. The editing by Pat Harris made the writing better. And thanks to Stephanye Hunter, my editor at the University Press of Florida. I've worked with a lot of great ones. She's one of the best.

PROLOGUE

It wasn't the first dinosaur I'd crossed in my life, but it was the first I'd ever engaged in eye-to-eye combat with. It glared up at me from the white sand where it stood its ground, scaly black toes and three-pointed claws digging into the white sand as if itching for me to come closer. The cloaking on the nape of its neck momentarily flared. Was I a friend or foe? Would it fight or take flight? Then, it stretched its neck forward, pivoted its head, and slightly opened its snout-like beak as if in gestures of appeasement. It flipped its tail up and down again and waited—patiently, fearlessly. Then I made my move.

I slowly reached into my pocket, pulled out a peanut, extended my palm, and waited. Now it was his turn. With a swoosh of feathery forearms, it lifted from the ground, propelled itself forward, and landed with pinpoint precision on my hand. The peanut was gone. So was the Florida scrub-jay.

It was a remarkable feat of nimbleness for a cousin (almost a hundred million years removed) of the seven-ton *Tyrannosaurus rex.* Had I stared down that beast, it would have devoured me with a few chomps and left the peanut behind.

Modern birds such as the Florida scrub-jay are not *like* dinosaurs. They *are* dinosaurs—every bit as much as *T. rex,* a member of the dino group from which birds descended. *T. rex,* according to recent scientific evidence, may also have had feathers, not for flying, obviously, but for warmth, camouflage, or display.

Like some of their dinosaur ancestors, modern birds build nests and lay eggs. Their feathers grow from the same type of cell as do the plates of alligators and crocodiles, their closest living ancestors, which also share their high-efficiency respiratory system. The long, dinosauric neck of modern birds, while usually hidden by feathers, is startlingly apparent when one pulls the turkey neck from the cavity of the Thanksgiving bird. If you want to see the dinosaur in the bird, look at one without feathers.

Birds like the Florida scrub-jay remind us that dinosaurs are not extinct. Those ancestors of modern birds squeaked through the calamity that killed many of their closely related cousins, including all of the more primitive birds with long tails and teeth, ones that had developed wings on all four legs instead of just their forearms, and those whose wings consisted of skin stretched between their limbs.

Once the lucky few threaded the needle of geologic catastrophe to make their way into the Cretaceous period, they didn't just survive. Eventually, they would prevail. Today, the ten thousand species of bird-dinosaurs compose one of the richest assemblages of any animal.

Their success was by no means a given. They traveled a long and treacherous evolutionary road to get where they are today—a road that began, in one sense, in the Middle Triassic period, 240 million years ago. Then, dinosaurs roamed parts of the megacontinent known as Pangaea, a single landmass that covered much of Earth and was surrounded by an ocean. There, small, fleet-footed theropods lived among their larger dinosaur cousins, including *T. rex.*

These hen-size theropods may have scurried about rocks, waded through ferns in cycad forests, and built nests in hollow conifer logs. Other contemporaneous dinosaurs may have climbed trees with their large claws and used their wide forearms to glide back to the ground. Some theropods had longish snouts with teeth for catching fish, lizards, or other prey, while the ground-dwellers may have had toothless beaks better suited to collecting, cracking, and eating seeds than to catching live prey.

Over the next 150 million years, as the landmasses slowly drifted apart on their way to forming what today are the continents, new dinosaur species emerged, some with distinctly birdlike features such as wings. They lived among a fantastic assemblage of plants and animals.

Then, sixty-six million years ago, this rich veneer of evolving life on Earth was shattered and the course of evolution changed. It all happened within a matter of minutes when a space rock, about as wide as Mount Everest is tall, pierced Earth's atmosphere and slammed into the coast of Mexico's Yucatán Peninsula at forty thousand miles per hour. The asteroid punched a hole in the ground a mile deep and more than a hundred miles wide. The impact occurred about six hundred miles from what is today the Florida peninsula. Back then, during a time of naturally high sea level, "Florida" would have been little more than a few small islands in a shallow, sunlit sea.

The asteroid impact released a heat pulse that ignited trees, shrubs, and grasses hundreds of miles away. The force ejected rocks and other debris into a low orbit around Earth, which slowly rained down in fiery red paths, igniting wildfires around the globe. Much of North America, which lay directly in the path of destruction, was ablaze within minutes. The landmasses to the south were also shaken but may have escaped the worst.

Vaporized rocks and sediment from the crater spewed poisonous sulfur into the air, and soot from burning debris began to block the sun's rays until day became night. The atmosphere cooled, and photosynthesis stalled. Without sunlight, many food webs collapsed. Three-quarters of all species eventually perished.

Many postimpact fossils suggest that of the myriad kinds of dinosaurs living at the time of the asteroid impact, only one small group survived, perhaps living on a cluster of ancient southern continents. Their forearms were covered with feathers, in some species more like quills than plumage. Some of these avian dinosaurs developed retractable legs by making the femur—the upper long bone of the leg—horizontal. This set the evolutionary stage so they could fold up their legs for streamlined flight but extend their feet for landing.

Some developed light, hollow bones and a compact skeleton. Rather than using true lungs, these early birds relied on air sacs throughout their bodies. This novel breathing system provided a continuous flow of oxygen during both inhalation and exhalation. Although this may have been an adaptation to lower oxygen levels in the atmosphere during the Cretaceous, the system could also meet the high oxygen and metabolic demands of energy bursts required for flight.

As the skies slowly cleared above the smoldering planet, the surviving dinosaurs faced a dire world. Because forests around the globe had been burned or flattened, insectivorous tree-dwelling theropods may have quickly died off while the seed-eating ground-dwellers prevailed, according to Daniel J. Field of the Milner Centre for Evolution at the University of Bath in England. Seeds could have remained viable for fifty years in some places and outlasted the worst of the darkness, fires, acid rain, plunging temperatures, and collapse of photosynthesis. This may have allowed the postapocalypse ground-dwellers to inherit the Earth. Or perhaps those that flew best, fared best.

Whatever their secrets of survival, these early ancestors of modern birds may have arisen on West Gondwana, the clustered landmass of South America and parts of Antarctica, or at least it was from there that their flying descendants dispersed across Earth. According to Santiago Claramunt, associate curator of ornithology at the Royal Ontario Museum in Toronto, some of these early birds, such as the ancestors of eagles, owls, and woodpeckers, eventually crossed to North America across a land bridge that joined these two continents after the global catastrophe. Others, such as the ancestor of songbirds, reached the Australian continent through Antarctica, which was still connected to both continents and was covered by forest.

Songbirds, distinguished by their highly developed vocal organs, diversified in Australia. From there they slowly colonized Eurasia, where they continued to evolve before eventually reaching North

America. Among the New World arrivals was the ancestor of the Corvidae—the family that would eventually give rise to crows, ravens, rooks, jackdaws, magpies, and jays.

Jays—a genus, or branch, of the crow family—share many characteristics of crows. They tend to be gregarious, with strong social organizations, strong pair-bonds, and males that help to build the nests. People often comment on the curious, intelligent, playful, and, at times, pugnacious personalities common to crows and jays.

These New World jays eventually split into their subgroups. Among these is a branch of the family that evolved millions of years ago to live in the hot, dry landscape of western North America. This landscape was likely composed of dense thickets of dwarf trees, including oaks or other scrub. It probably resembled the hot, low-growing, fire-prone habitats today known in parts of the West as chaparral. The word denotes any landscape of dense shrubs or dwarf trees. The New World jays that inherited this western habitat would become scrub-jays.

High sea level at the time had flooded much of the Gulf Coast and pushed water far north into the Mississippi River basin, creating vast deltas and effectively dividing the continent. This prevented many western species from reaching the East. But that began to change about two million years ago as sea levels fell. The lowering water not only shrank the deltas but even exposed the sandy continental shelf along the Gulf of Mexico that connected Mexico and the Southwest directly to the Florida peninsula. Where once there were wide deltas, now there was a dry land bridge. As the climate further changed, a western habitat expanded eastward over dunes left by the falling seas, and some western species began to disperse slowly eastward. Among this slow-moving caravan of expanding species was the western scrub-jay. But it did not travel alone. Accompanying it eastward were a tortoise, a horsefly, a sand-burrowing cockroach, a prickly pear cactus, and several other westerners.

All of these species eventually reached Florida, where they joined the riot of life thriving there. For thousands of years, the scrub-jay was just an extension of the original population and differed little from its western counterpart, living in a landscape that was an extension of the western chaparral.

Thousands of years after the scrub-jay arrived in Florida, sea level in the Gulf of Mexico began to rise again. (It rose and fell many times over millions of years.) High seas once again flooded the land bridge across the Gulf and severed the scrub-jay from its western origins. The bird and other species that had traveled with it were permanently cut off from the ancestral populations. That's why biologists call the Florida scrub-jay and its compatriots "relict" species, a word related to the Latin "widow." Isolated in Florida, the bird's evolution began to diverge. As sea level continued to rise, the bird became more isolated still. As seawater flooded the low-lying Florida peninsula, scrub-jays were limited to high dunes and sand ridges deposited along former coastlines from high seas in earlier geologic epochs. These places where the scrub-jays survived are called "refugia," related to the Latin "flee." The scrub-jay, earlier widowed from its ancestral population, had now become a refugee in its own land.

Over time, different populations of the birds became mainly concentrated in four sandy upland regions of the Florida peninsula: the Atlantic Coastal Ridge on the East Coast; the alluvial dunes along the Gulf of Mexico in southwest Florida; the Central Ridge; and in regions around what is today the Ocala National Forest.

Given the Florida scrub-jay's precarious status as a widowed refugee, it's a double miracle the bird survives at all. Its ancestors were the only dinosaurs to survive the meteor impact sixty-six million years ago. Having made an improbable journey across North America two million years ago, the bird then became confined to isolated patches of habitat on a peninsula otherwise hostile to it.

But today, the scrub-jay faces its greatest challenge to survival. *Homo sapiens* arrived on the Florida peninsula only about fifteen thousand years ago. By then, the scrub-jay had been part of the Florida landscape for perhaps two million years. The arrival of rational man changed everything. Humans were the first species to bring a knife to the fight for survival. They were the first to eventually amass the destructive power of an asteroid. There has not been a major mass extinction since the Chicxulub rock struck sixty-six million years ago—until now, as scientists document the beginning of the next mass extinguishing of species on Earth. As the fiery asteroid defined the end of one geologic age, the current wave of annihilation marks the grand opening of the next: the Anthropocene epoch.

In many ways, fire in one form or another has defined the very existence of the Florida scrub-jay. Its earliest ancestors narrowly escaped the conflagration that followed the asteroid impact. What's more, for millions of years, the Florida scrub-jay's needs and behaviors have been intertwined with fire. They have been consigned to an early-stage habitat that depends on periodic lightning fires to clear away dense, predator-prone older growth and promote new and younger oak brush. The birds have evolved so closely and for so long with fire that no sooner does the smoke clear from a controlled burn than scrub-jays return to forage in the devastation. Surely, a faint scent of ancient genetic memory must be calling them home.

Luck, resourcefulness, and tenacity enabled the Florida scrub-jay's ancestors to survive sixty-six million years ago. Whether that will carry them through the impact of the Anthropocene remains to be seen.

PART ONE

THE ATLANTIC COAST

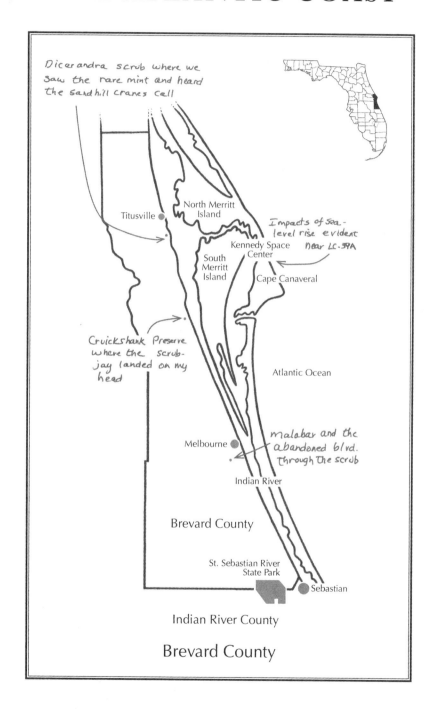

Dicerandra scrub where we saw the rare mint and heard the sandhill cranes call

North Merritt Island

Titusville

Impacts of sea-level rise evident near LC-39A

Kennedy Space Center

South Merritt Island

Cape Canaveral

Cruickshank Preserve where the scrub-jay landed on my head

Atlantic Ocean

Melbourne

Malabar and the abandoned blvd. through the scrub

Indian River

Brevard County

St. Sebastian River State Park

Sebastian

Indian River County

Brevard County

I

BREVARD COUNTY, 1925

Almost a century ago, the road my grandfather Freddie traveled from his home in the town of Sebastian, Florida, was mostly a rutted, tire-busting, sand- and coquina-packed byway running along a dry ridge next to the Indian River. The newly completed Dixie Highway had recently joined the railroad and riverboats as a way of traveling along the coast of St. Lucie and Brevard Counties.

Freddie apparently liked to take rides along the Dixie Highway, partly for relaxation but also to show off his new car. He was among the first residents in Sebastian to own one. His Overland 6 was note-worthy enough that the *Vero Press,* in nearby Vero, described it as "one of the prettiest sedans among our automobile population." With his new car, dashing mustache, dark eyes, and head of thick, dark hair tossed back, Freddie must have been a sight: a modern self-made man driving through the latest edition of paradise.

Sebastian is situated on the west shore of the Indian River. A distant barrier island forms the east bank, and on the other side of this lie the beaches of the Atlantic Ocean. The Sebastian environs have always been known for their pleasant climate and sweeping river view. In Freddie's day, the area was also so rich in birds that many of the most famous ornithologists and naturalists of the day visited there, and the beautiful coastline had begun to attract sportsmen, birders, vacationers, and wealthy retirees from up north.

On cool mornings, breezy afternoons, and mild winter days, Freddie's drive along the Dixie Highway brought balmy breezes across his cheeks and heavenly views of the Indian River's mile-wide waters luffing with sailboats and a few puffing laggards of the steamboat era. Pelicans rode breezes along the river with occasional flaps of their wings, roseate spoonbills waded the flats, and various kinds of gulls congregated along the sandbars.

Freddie had arrived in Sebastian around 1916, during a time of fast and perilous change. The railroad along the river, completed about two decades before, had brought new settlers into St. Lucie and Brevard Counties. By the early 1920s, demand for land had sent prices soaring. With the last stretches of the Dixie Highway completed, the iron horse of the railway had given way to herds of metal ponies along the highway. Swamps were drained, citrus groves planted, towns invented, named, and renamed.

Moving from the less affluent Washington County, Georgia, Freddie had come to Florida in search of better living. Like many new arrivals, he was probably also drawn to the area by magazine and newspaper ads describing the idyllic life there, a place where the infirm came to recuperate and the healthy apparently never got sick. A 1916 city directory described Sebastian as being "located 215 miles south of Jacksonville in the center of the famous Indian River fruit belt. . . . It is on a ridge overlooking the Indian River, and is but 3 miles from one of the highest points between Jacksonville and Miami, the health conditions are ideal. Good schools and churches. A community of live, wide-awake people." Fewer than two hundred lived there at the time.

Freddie, a machinist, set up an auto-repair garage at the intersection of the Dixie Highway and Sebastian's main street in a large, white, metal-sided building with a false front bearing his last name, Walters, in bold, black letters. He taught many of the wealthier residents how to drive their new cars. He lived with his wife and six children in a two-story house at the corner of Palmetto and Louisiana Avenues, a five-minute walk along Main Street to his garage. Although he didn't truly own his Overland 6 yet—the bank did—the car signaled his newfound prosperity, and the Dixie Highway, which connected a growing labyrinth of roads to the East and the Midwest, must have seemed like the on-ramp to a dream.

Occasionally accompanied by friends or his young sons—including my father— but more often alone, he would drive up to Melbourne, cross the wooden bridge to the barrier island, and go fishing. Sometimes he ventured as far north as Merritt Island, crossing the new bridge at Cocoa. He drove through the citrus groves that covered the southern part of the island and sometimes returned home with crates of oranges or grapefruit in the back seat. Sometimes he braved the sandy roads of North Merritt Island, where only a few fishermen or hunters had built outposts in the surrounding scrub. Occasionally Freddie drove all the way up to a big hardware store in the booming metropolis of Titusville, seventy miles north of Sebastian, for tools or other supplies. Traveling up to Titusville wasn't a banal workaday drive. In those days, especially in a fancy automobile, it was a journey worth telling your children and grandchildren about.

As Freddie drove along the coast, he passed the scattered river settlements of Roseland, Micco, and Malabar surrounded by a sea of pines, sabal palms, or low oaks along the river. Continuing north toward Melbourne, the road passed along low, sandy dunes where tropical morning glories twined across the tops of the rises. Along stretches, roadsides bloomed with yellow or orange sunburst blanketflowers. In places, the shiny, round leaves of sea grapes hung from gnarled gray trunks along the road.

Where the highway veered inland from the river, the roadsides were thick with tough oak bushes and shrubs and graced by a few tall, skinny pines. In spots, the thick walls of scrub rose up to the windows on both sides of his Overland. Now and then, these thick barricades of branches and palmettos would yield to younger, more recently burned scrub and natural patches of bare white sand. Where the road dipped and the sand became moister, the scrub dissipated, and thicker stands of pine rose against the blue sky. This ebbing and flowing balance of pine and scrub stretched along the entire 150-mile length of what was known as the Florida Atlantic Ridge.

In mornings as the Overland 6 approached, scrub-jays would peer from perches near the clearings and dive into the scrub as he passed by. Sometimes they would dart across the highway or flit along the roadside as their sentinels sounded a single, high-pitched *Kweep!* from pine snags in the distance. More than likely, Freddie called them "jaybirds," or at least that's how my father referred to them, and it would make sense that he had inherited the name from his father.

Although a kind of jay, they differed from the ones Freddie had known back in Georgia. The Florida bird seemed more streamlined, with a longer, graceful tail, and it lacked the crest and black chin-strap of the blue jay. Whereas the Florida jay's hues faded from one to the next, the Georgia bird had bold black and white wing bars. If the Georgia bird was bold in appearance, the Florida jay was understated, even princely. Its mostly blue cape gave it a royal air. The bird could be skittish at first but grew accustomed to people more easily than most of the standoffish blue jays in Georgia.

In the scrub clearings, the Florida bird seemed perfectly at home on the bare ground, hopping through the white sand, picking up and caching acorns just below the surface, or prospecting for lizards. The birds deftly navigated around the bases of the scattered bushes, pivoting their heads as they searched for crickets, grasshoppers, or caterpillars in the vegetation. Then, with a few labored wing strokes, they would fly up to a branch of a tree or a bush. In flight, their bodies seemed just a little heavy for the size and lift provided by their wings. Perched, a scrub-jay hung its tail straight down, with its neck extended as if on the lookout or, if resting, with its neck pulled down closer to the body. At one moment, a bird might preen itself. In the next moment, it could snap its head up and zero in on another scrub-jay invading its territory.

After a few miles of travel through the walls of thick oak scrub along the highway, the road swung east back toward the Indian River, and Freddie could see Pelican Island and, beyond that, across two miles of water to the green horizon of the barrier island. Unlike the subdued scrub, the river always teemed with an acrobatic show of wheeling terns and gulls, egrets, herons, pelicans, and phalanxes of small waders in the shallows. Ospreys dive-bombed fish, and slender terns dipped for minnows. It made life seem easy.

Freddie had moved to Sebastian just in time to witness a renaissance of shorebirds in the region. A mere two decades earlier, many once-common ones along the Indian River had been scarce. During the Victorian era, birds were slaughtered and their feathers used to adorn women's hats. Great egrets and snowy egrets were especially highly prized for their long, lacy breeding plumage. It took an odd leap of imagination—or perhaps a habit of blindness—to believe that feathers bestowing such beauty on birds could transfer their majesty to a species that ruthlessly hunted them.

Pelicans were also shot for their attractive feathers. Beliefs that the birds competed with fishermen also marked them for slaughter. By 1900, Sebastian's Pelican Island had become the last brown pelican rookery on Florida's east coast. Fearing the bird's extinction, in 1903, President Theodore Roosevelt established the island as a preserve. This, along with new laws protecting birds, had by the 1920s helped many species to thrive along the river once again.

Known for marvelous birdlife, St. Lucie and Brevard Counties had attracted notable ornithologists and botanists from around the country, including Frank M. Chapman, the curator of ornithology at the American Museum of Natural History, and John Kunkel Small, a curator of the New York Botanical Garden. Their accounts, along with the naming of Pelican Island as the first national bird preserve, helped to solidify the area's reputation among naturalists of the early and mid-1900s as a paradise redeemed.

The river took center stage for ornithologists, while the counties' wetter pinewoods and marshes drew notable botanists. Many viewed the hot scrub on high, drier land—seemingly nondescript except for its abundance of jays—as an almost alien habitat. As one researcher wrote: "The vegetation is mostly dwarfed, gnarled and crooked, and presents a tangled, scraggly aspect. It appears to desire to display the result of the misery through which it has passed and is passing in its solution of life's grim riddle. . . . Here the sun sheds its glare and takes its toll on the unfit."

Less than two miles north of Freddie's home, the Dixie Highway crossed the Sebastian River, and as the wooden planks clapped under his tires, Freddie could see numerous birds flapping along the green banks, above the torpid currents, and across a landscape of almost unimaginable extravagance and grace. Chapman described it as "a beautiful river; no words of mine can adequately describe it. . . . [T]he banks as we found them are . . . densely grown with palms and cypresses which, arching, meet overhead, forming most enchanting vistas, and in many places there was a wild profusion of blooming convolvulus and moon-flower."

But like the Indian River, into which it emptied, the Sebastian flowed with its own macabre history of avian destruction. Not far from where Freddie crossed, Chapman had earlier discovered the last known roost of the Carolina parakeet. Word quickly spread around the small town, and a few days later, another man showed up at the site, placed a gunnysack over the tree opening, and captured all the birds. He then cut down the tree and sold it to the University of Kansas.

The Carolina parakeet became extinct shortly afterward, but the birds were rumored to have lived along the river into Freddie's day. Specimen collectors, some working on bounties, still hunted them, as they did the last ivory-billed woodpeckers. During the same period, the passenger pigeon, also once common in parts of Florida, became extinct.

The extinction of so many well-known species played out in the local and national press at the time, in talks at Audubon Society meetings, and in library gatherings. In 1923, an article in the *New York Times* summed up the growing national concern: "The question of wild life conservation has reached a crisis . . . with certain societies and groups pitted against sportsmen's organizations in perpetual altercation as to what species shall be saved from extinction."

After paralleling the bird-rich river shore, just north of Valkaria, the Dixie Highway turned inland, and once again Freddie found road lined by mile after mile of scrub and jaybirds. Their small feathers mostly useless for the millinery trade, the birds were left alone in their secluded oak thickets, living in a world apart from people—except for a few passing cars—and focusing on hawks, snakes, and other age-old enemies. Keeping to their tight family groups of six or eight birds, they didn't coalesce into noteworthy flocks but tenaciously defended their territories from neighboring families. While their large territories gave the impression that the birds were scarce, they owned nearly every square foot of scrub in sight. At the time, there were thousands of territories around the county claimed by seven or eight thousand birds.

In the same way the scrub was largely overlooked by botanists and ornithologists at the time, developers also expressed little interest in it. They were busy buying and selling the pricey coastal lands. If the scrub was noticed at all, it was mostly by citrus growers who had discovered that orange and grapefruit trees, if fertilized and irrigated, would thrive on land cleared of scrub. Elsewhere, settlers were converting marshland to agriculture by digging drainage canals. Ten miles west of Freddie's home, one canal-building enterprise was so big it gave rise to the new town of Fellsmere, a place Freddie sometimes drove to along the road that ran alongside a spur off the main railroad.

In 1924, the curator John Kunkel Small prospected for plants in the area immediately around Sebastian. He described the region as "uninhabited except for an apparently wholly out-of-place settlement . . . of Fellsmere." Unlike many naturalists at the time, Small praised the subtle beauty of the scrub but had little to say about the vast area of oak thickets around Sebastian, which held one of the county's largest concentrations of scrub-jays. This was probably because he was so taken by the sprawling garden he found lower down on the ridge, which he described "as prolific a natural flower garden as I have seen in Florida, for here the pine-woods were a mass of flowering plants."

About fifteen miles north of Sebastian, Freddie entered the town of Tillman and crossed the bridge over Turkey Creek, which, like the Sebastian River ten miles back, also flowed into the Indian River. The town's name had recently been changed to the more alluring Palm Bay. Scrub crowded the bluffs along Turkey Creek, according to reports from the time. Where there was scrub, there were scrub-jays. If Brevard and St. Lucie Counties had been called the "Scrub-Jay Coast," no one would have asked why.

Despite a few growing settlements, scattered orange groves, a railroad, and drainage projects to convert marshland, Brevard and St. Lucie Counties were still mostly open range. Lightning fires still regularly raged across a fenceless, largely roadless landscape as they had for thousands of years. Smokey the Bear hadn't yet been born, and the unchallenged fires burned off the old scrub and permitted the constant renewal of the newer growth that scrub-jays needed for survival.

But things were already rapidly changing. Massive land-reclamation and canal-building projects like those happening at Fellsmere increasingly pierced the heartlands of Brevard and St. Lucie Counties. Small groves of citrus were giving way to industrial cultivation, which began to claim more and more of the dry uplands. As more people moved into the region, the once-overlooked scrub was becoming prime real estate for houses and settlements.

During the earlier and bloody era of plume hunting, citizens had fought vocally for the preservation of birds. They witnessed the savagery, and their outrage led to new conservation laws. But the new age of commercial development was different. The pain inflicted on animals by modifying or obliterating entire landscapes was mostly hidden. When developers drained a marsh for crops or cleared scrub for citrus, people saw only progress and cheered—even as birds and other species suffered out of sight.

But by the mid-1920s, the economic boom that had brought Freddie to Florida was stalling. Land had become so expensive that sellers could no longer find enough buyers. Businesses went bankrupt. Towns and counties that had overextended themselves financially on road-building or swamp-draining projects defaulted. Banks stopped lending. People who had lived lives of relative leisure became gripped in intolerable fear and stress as they lost their jobs, homes, and cars. Freddie lost even more in 1927, when his wife, who was barely thirty, died suddenly from a brain hemorrhage. Freddie moved up to

Melbourne before eventually pulling up stakes in Florida altogether and returning to his Georgia roots. Several of his children, including my father, remained behind in Sebastian to pretty much fend for themselves through the Great Depression. But with all the suffering and chaos consuming the human-built world, life in the scrub went on as if nothing at all had even happened.

2

THE SCRUB WHISPERER

History is always a work in progress. More than a record of what happened, it's the piled-up commentary about how and why it happened. So, nearly a century after Freddie traveled the Dixie Highway, I decided to make the drive myself. Not in an Overland 6, of course, or traveling a bumpy road. I would make it in an air-conditioned Toyota minivan with a stereo radio, an automatic transmission, and at fifty miles per hour along a multilane highway, not to mention while maintaining communication with an array of orbiting satellites just to make sure I wasn't inconvenienced by a wrong turn. By keeping on the seat next to me the 1926 edition of *Mixer's Road Guide and Strip Maps* to help me more or less trace Freddie's route along the jaybird coast, I hoped to witness a glimmer of what he saw and better wrap my mind around the paradox we call progress. The scrub-jay is all but gone from mainland Brevard County. Determined to bring Freddie with me along what is now called the Space Coast, I tucked an old photograph of him between the dashboard and the windshield of my car, facing outward so he could see how the county had changed. I would be Freddie's guide, since he wouldn't know where he was anymore, and then I would find one for myself.

So I called Paul Schmalzer, a plant ecologist in Titusville who worked on contract for the National Aeronautics and Space Administration (NASA). Schmalzer had published more than seventy scientific papers—many of them on the ecology of Brevard County—since receiving his PhD from the University of Tennessee, Knoxville, in 1982. He probably knew more about Brevard County scrub than anyone. He knew a lot about the history and a great deal about the scrub-jay, too.

Cheery and friendly, Schmalzer was enthusiastic when I met him at the Enchanted Forest Sanctuary in Titusville early one morning in 2016. He thought it would be a good place to begin because the preserve held one of the few remaining areas of scrub—all twenty-five acres—near the town. We parked our cars at the visitors' center, where we met the preserve's land manager, Xavier De Seguin Des Hons, who would accompany us on the trail through the 470-acre Enchanted Forest.

As Schmalzer, Des Hons, and I walked from the visitors' center, the gravel path passed beneath some sunny palms and through palmettos before gradually descending into a gloomy grove of massive oaks draped in Spanish moss. Smaller laurel oaks, red maples, black gums, sweet bays, and sugarberries shaded an understory of holly and marlberry. Large ferns clustered around the tree trunks. I almost expected to see someone in a loincloth or maybe a shy couple wearing fig leaves step into the open. The grove carried all the storied romance of an ancient forest, and it was from this enchanting hammock the preserve took its name. For all the world, the sight evoked Henry Wadsworth Longfellow's poem *Evangeline:*

> This is the forest primeval.
> The murmuring pines and the hemlocks,
> Bearded with moss, and in garments green, indistinct in the
>   twilight,
> Stand like Druids of eld, with voices sad and prophetic.

The forest was old but not ancient, and it was far from pristine. The original growth had long since been logged. While some visitors might be tempted to describe the forest of large live oaks and draping moss in purplish prose, like I almost just did, ecologists usually resort to military analogies to describe what's going on there, using words

like "war," "battle," or "combat." As Schmalzer described the native gi-
ant air plants (*Tillandsia utriculata*) that once festooned the trees like
huge pineapple tops, he held his hands a shoulder's width apart. "The
Mexican bromeliad weevil—the evil weevil—was inadvertently imported from Veracruz, Mexico, to Florida by a Ft. Lauderdale bromeliad nursery," he said. By 2007, it was attacking the vulnerable species
at the Enchanted Forest. Des Hons added that the massive air plants
used to hang everywhere from the big trees. Besides the evil weevil,
armies of exotic plants had established beachheads in the preserve. In
the 1980s, imported monkeys roamed the area after being cut loose
from the nearby Tropical Wonderland theme park when it closed.
The Enchanted Forest was beginning to sound more like a haunted
one.

For all the Enchanted Forest's ecological disruptions, its condition
has vastly improved in recent decades. The monkeys are gone. Repeated burnings have restored some of the scrub; volunteers and staff
combat invasive plants; and experts such as Schmalzer promote the
sanctuary's educational and ecological value. For numerous visitors
and students, the Enchanted Forest is a living textbook on both the
beauty and sorrow of Florida ecology. And when you have a scientist
like Schmalzer with you, the textbook becomes available in an Au-
dible format.

The trail ascended gradually as we left the hammock, and Schmalzer looked back. "The ground is getting drier from where we were,"
he commented, explaining that we had just left a mesic, or moder-
ately dry, hammock, where the water table is typically just below the
surface. (By contrast, in a hydric, or wet, hammock, the water table
is at the surface much of the time.) "Now we're maybe six inches
higher above the water table. Look how the trees are changing." He
pointed out how pines were more common than the big hardwoods
and how the sandy soil had lightened in color. We climbed the hill.
"We're moving into the xeric, or dry forest, several feet above the water table," he said, pointing out that palmettos now dominated. At the
top of the incline, we reached scrub-jay territory—just as the Florida
ecology textbooks tell you.

"Here's the scrub," Schmalzer said, spreading his arms as if ready to reach out and pick up all twenty-five acres. Now, about six feet above the water table, the sand had become bright white, the ground bone-dry. But the scrub we encountered wasn't the waist-high thicket described in the books I'd read. It was a charred landscape worthy of its own meteor strike. Des Hons said he'd intentionally burned the parcel several months earlier.

"Just what the doctor ordered," Schmalzer said, explaining that the burn didn't make it any less scrub but returned it to an earlier stage of succession. The real fountain of youth in Florida is a fountain of flames. It had burned off the top of the scrub—given it a buzz cut—but the roots and millions of fire-resistant seeds remained on or beneath the blackened sand-like stem cells ready to re-create what they were meant to be. The dry weight of oak roots in scrub can exceed that of the tree parts aboveground. Although the scene looked catastrophic, for the plants and animals it was a perfect picture of hope. It had been a cleansing fire. Without it, the scrub would have become an overgrown forest and forced the scrub-jays to abandon it. That is, if the intense fragmentation of the scrub around Titusville hadn't caused them to abandon it long before.

Schmalzer mentioned how the Florida scrub likely resembles the habitat where the bird's western ancestors lived millions of years ago and where closely related species of scrub-jays still live today. The chaparral, as it's called in parts of the West, has a similar plant-structure with low-growing fire-resistant species, although the species themselves are different. Western scrub-jays also depend on periodic fire for habitat renewal. Although Florida scrub and western chaparral bear different names today, the term "chaparral" comes from a mid-nineteenth-century Spanish term meaning "dwarf evergreen oak," which describes Florida scrub to a tee. *Merriam-Webster's Collegiate Dictionary* defines chaparral as "a thicket of dwarf evergreen oaks" or, more broadly, "a dense impenetrable thicket of shrubs or dwarf trees." You couldn't find a more accurate description of the Florida scrub if you tried. What's more, both scrub and chaparral are shaped by alternating drought and rainy seasons. The identical historical meanings and close ecological relationships of "chaparral" and "scrub" would lead one to reasonably conclude that they are different names for the same thing.

After this brief detour in our conversation, I asked Des Hons where the scrub-jays had gone when the Enchanted Forest scrub was on fire. He said they hadn't nested here for a quarter century. Five years had passed since the last blue-feathered ghost had been spotted passing through. With that, Des Hons politely changed the subject.

"It's all coming back," he said, leaning over to point to the bright-green sprouts on the charred palmettos and peeping from the blackened oak stems near the ground. In better times, scrub-jays would already have been returning to forage for insects in the treeless landscape. In another three or four years, the scrub-jays would have begun nesting there again. A new cycle would have begun and would pause when the next lightning strike burned it to the ground a few decades later. Or, in this case, the next time Des Hons set a drip torch to it.

"Will the scrub-jays ever be back?" I asked.

"No," Schmalzer said, stepping into the conversation. "It's not big enough, and it's surrounded by forests scrub-jays won't go through. It's just a lonely island of scrub." He said scrub-jays need archipelagos—a series of patches over a large area. This patch was barely big enough for a single scrub-jay territory. He called it "a museum piece." Still, regular burning maintained the patch for other rare scrub species, such as the gopher tortoise. He pointed out the sand-ringed mouth of a burrow on the burnt ground.

"They hide there during fires and come out just a few hours afterward," Des Hons added. "A lot of other species use the burrow, over a hundred of them, including the eastern indigo snake. Without the tortoise burrows, many of these other species wouldn't be here. It's called a keystone species. Small patches are worth preserving, even if scrub-jays don't live here." It is the same idea, I suppose, behind art museums displaying the fragments of once-great sculptures and mosaics.

But ecologically, the small burned rectangle of scrub where we stood in the Enchanted Forest was no longer even part of a mosaic.

It was a sad single tile—a largely nonfunctional disembodied piece of the former whole. As in a work of Romanesque art, all the tiles in a mosaic work together—or they may not really work at all.

One thing I learned during our visit to the Enchanted Forest is that you can't talk to knowledgeable people about the scrub-jay without talking about fire. And that usually puts the chat on a pleasantly slippery slope about ecology because you can't easily talk about fire without mentioning the pinewoods that often border the scrub, or, dare I say, chaparral. Scrub, with its fire-resistant plants and scarcity of leaf-litter or kindling, is slow to ignite. It burns periodically but hesitantly. The sap-laden pinewoods that border scrub, however, are always up for a good burn. From these fire-ready woods, which are often slightly lower than the scrub, the stoked flames take the hill like a flamethrower and provide the heat and oomph to ignite the fire-resistant scrub. The pinewoods need to burn for the scrub-jays' sake but also for their own good. Otherwise, they get tall, thick, and overgrown and, to bring it back to the bird, become barriers between the archipelago of scrub.

Historically, lightning fires in Florida broke out during spring and early summer storms, when the plants grew most and could recover quickly. Fire often resulted from "dry thunderstorms," where there was lots of lightning but the precipitation evaporated before it hit the ground. Unlike prescribed burns used by forest managers—and that Des Hons had set a few months before our visit—wildfires left an unpredictable wake of life-giving destruction. They didn't burn in straight lines or rectangles. Local hydrology, wind, topography, and firebreaks, such as ponds and lakes, made them zig and zag, smolder or die down, and then suddenly flare up. Flames gingerly crept through some areas and roared across others, splitting into Y-shaped arms around ponds, wet areas, and other obstructions and leaving unburned tails until the flames rejoined. It is in these less frequently burned fire-shadows that scrub often occurs. In the end, natural fires leave behind one big, beautiful mess that ecologists call a mosaic and that scrub-jays call nirvana. Unlike a homogeneous scrub, a mosaic of different-aged oaks guaranteed a supply of acorns for scrub-jays and a diversified menu of insects dependent on plants at various stages. The irregular, stop-and-start-again pattern of natural fires also made for some unburned or lightly burned areas, ensuring that scrub-jays would have a few quarters to live in even when the rest of their housing had burned.

Des Hons motioned us along. We gently descended the ancient dune and scrub and came to a creek—the wreckage of an earlier-century industrialist's dream of building a canal to drain the marshes in the region. But where the dreamer had expected to plow easily through soft dunes, he encountered instead a nightmare of buried rock—a cement-like mixture of packed coquina shells—the geological foundation of the Atlantic Coastal Ridge that had been built by deposits of surf pounding on small shells over thousands of years. Mountains of blowing sand gradually accumulated on this cement-like core, forming a high, dry ridge along the coast that Freddie had driven on a century ago and that Schmalzer and I followed. Having descended from the scrub, we were encountering some of the exposed bones of the ridge, just as the steam shovels had several decades earlier.

"I guess they broke quite a few buckets on this stuff and finally decided to call it quits," Des Hons said as he reached up to grip a piece of exposed caprock on the bank above the would-be canal.

"This makes the scrub on the ridge unique," Schmalzer added. "Most scrub occurs on deeper sand. Certain species grow here and not in other places." He reached out to touch the bright-green fronds of a small caprock fern clinging to a cleft in the exposed coquina. He said it was specifically adapted to the outcroppings of the Atlantic Coastal Ridge.

As we turned around and headed back to the visitors' center, Schmalzer pointed past the trail to a waist-high array of bright-crimson flowers arranged concentrically around a straight stalk standing in the sun. "Coral bean. A native flower and hummingbird favorite," he said gleefully. A few steps farther along, he stopped and carefully pulled a fine thread from the edge of a cabbage palm leaf. "This is the scrub-jays' favorite material for lining their nests," he said. Next to it was what seemed like the identical leaves of a saw palmetto. Schmalzer said the fine threads are "one way to tell saw palmettos from cabbage palms." He said the saw palmetto's berries are the source of an herbal remedy for some prostate disorders. It's a very common remedy in health food stores." Who would have thought that "scrub-jay" and "prostate" would be used in the same sentence? It was the kind of connection that reinforced the conclusion of the writer and ecologist Aldo Leopold in *A Sand County Almanac*: "To keep every cog and wheel is the first precaution of intelligent tinkering." It is surely the first precaution of any serious effort to save the scrub-jay.

As we returned to the visitors' center, Des Hons said that about a mile and a half away lay another patch of scrub. Perhaps we'd see some scrub-jays there. He offered to have his assistant drive us there in the truck. Schmalzer and I gladly accepted.

*   *   *

Along the way, Freddie would undoubtedly have been dazzled as we drove past shopping centers and franchises—AAMCO, Sonic, Mattresses4LE$$, MetroPCS, Sprint, H&R Block, Wells Fargo, and Chick-fil-A—or Dino-fil-A, as a paleontologist might call it. We passed a small strip mall, and the asphalt turned to a hard gravel roadway paralleling a chain-link fence next to a housing development and a mobile home park. Just beyond that, the gravel road turned to rutted white sand, and the truck tires began to spin. "We could probably make it farther, but we don't want to get sand up in the transmission or engine," the driver said. But for efforts to save the patch of scrub, the strip malls we passed on our way here would have by now exploded beyond their current boundaries and into what is now a preserve.

As Schmalzer and I got out and walked into the scrub, he pointed to some disintegrating pressboard and other debris along the road. He said that we stood in a Florida Power & Light Company easement in a part of the preserve that wasn't controlled. "It's an open sanctuary. People dump here."

We walked past the power line and into the scrub, and I imagined this was the kind of chaparral Freddie would have seen along much of the Dixie Highway. Scrub-jays would have been everywhere. Tall stands of pines would have lined the horizon instead of the clay-tile roofs of a housing development. But it would be facile to pass off the dilemma with simple regret. People need places to live, and in most places people choose to live, something else is going to be—or already has been—pushed out of its home. And so the question is not just where people choose to live but *how* they will live there. Clearly, the problem hasn't yet been figured out here.

Schmalzer and I "pished" for jays with a rapid *shreep, shreep, shreep,* but none appeared. Unperturbed, he said, "Here's a kind of prairie clover," dropping to one knee in the sand and grasping the woody stem and narrow compound leaves. "Won't be long before it blooms with small bluish-purple flowers. It's a member of the pea family." He said it was rare, although not listed, and grows only in scrub or scrubby flatwoods. I kept scanning the horizon for scrub-jays.

To spend a day with someone as curious and knowledgeable as Paul Schmalzer is to learn that if you come to the scrub looking for nothing, you'll find everything. But if you come looking only for the scrub-jay, you're apt to see nothing. He was the observer. I felt like a sightseer.

He studied the clover for a few moments and then leaned closer. "*Dicerandra* mint!" he exclaimed. There, growing near the road next to the clover, was the namesake of the Dicerandra Scrub Sanctuary— the delicate, diminutive *Dicerandra* mint, or Titusville balm. Mostly past its bloom, the mint still held a few blossoms from its November glory. Schmalzer said the plant grew naturally nowhere outside of Brevard County. And there it was mostly limited to a narrow range beginning near the Dicerandra Scrub Sanctuary to about thirteen miles north. Many of the surviving wild plants were on private land.

The plants' longish pink stamens, extending horizontally far beyond the petals, bore bright-rose striations and dots and were capped with tiny white finials shaped like traditional Dutch bonnets. The rare mint's soft white petals unfolded like two pairs of wings, one above the stamen and one pair below. The mint's small but stout, squarish stems held these delicate blossoms, three or four at a time, aloft for only a few weeks in fall.

Schmalzer rotated a stem between his thumb and index finger as he pointed out the small and delicate rosemary-shaped leaves. Even now, in December, the foliage was bright green. He pinched off a small leaf, crushed it between his thumb and index finger, and handed it to me. I detected the sweet fragrance before the crushed leaf reached my nostrils—a scent of lemon and clove, with a hint of rose. Schmalzer said beginners often confuse the plant with the similar prairie clover and that he himself had momentarily hesitated before identifying it.

Florida Today reporter Jim Waymer summed up the rare mint's predicament: "Tucked among tract housing and city wells is a patch of palmettos, scrub oak and bone-dry sugar sand, where a gangly inconspicuous herb of a minty variety sprouts up against all odds."

A bit like the scrub-jay.

Schmalzer and I both stood up when we thought we heard a scrub-jay call. We looked around. Nothing. Schmalzer continued to admire the rare balm. "The mint is very distinctive and beautiful, isn't it?"

As we walked a little farther into the scrub, Schmalzer pointed out that we were standing atop a ridge, a high point barely noticeable in the broader lay of the land. A sudden chatter of sandhill cranes sounded, not far away. Although we could not see them, their calls told us of a marsh or ephemeral pond just beyond the rise. Where we stood was bone-dry. But just a short distance away, down the invisible incline, cranes shouted for joy at the abundance of water. It felt like the scene from Marjorie Kinnan Rawlings's *The Yearling*: "Magic birds were dancing in a mystic marsh. The grass swayed with them, and the shallow waters, and the earth fluttered under them. The earth was dancing with the cranes, and the low sun, and the wind and sky."

We continued our walk. Still no scrub-jays. I tried to act like Schmalzer-the-Scrub-Whisperer but was determined to see some of the birds. "Oh, look over there," he said, pointing out two grapevine species shimmying up a large pine. He talked about the difference between Calusa and muscadine grapes and showed me how the underside of one's leaf had hairs, the other just a smooth, shiny surface.

There was nothing in the scrub that didn't fascinate him. A few steps farther, he pointed out a buckeye butterfly that kept rising and falling, landing on a small stone near another mint balm. Its tawny wings were striped with alternating light- and dark-brown striations, had spots that looked like eyes, and then flared to ochre orange at the bottom edge of the back wings. A pair of bright-orange epaulets stood out on each forewing. A bug flew out of the bushes, buzzed several revolutions around my head, and then dive-bombed into the cover of low oaks. The desert garden had come alive. Here, the sun may shed its glare and take its toll on the unfit, but those that made the cut have created a world of teeming beauty and wonder.

A few minutes later, Schmalzer pointed out a hog plum, or tallowwood. Not limited to the scrub, the dense shrub parasitizes the roots of nearby plants wherever it grows, he said. When it blooms in late spring, the yellow blossoms carry a heavy fragrance. Native Florida jam is sometimes made from their fruit.

After thirty minutes or so, we still hadn't seen or heard any scrubjays, and we walked toward the truck. "I know where we can go next," Schmalzer said. "There's good patch of scrub twenty miles to the south." He was talking about the Helen and Allan Cruickshank Sanctuary, in Rockledge.

The assistant dropped us off to pick up our car at the Enchanted Forest. Then Schmalzer, Freddie, and I headed out to our next stop, about forty-five minutes away, in the heartbreaking reenactment of Freddie's long-ago journey.

3

ISLAND-HOPPING

Fifteen minutes outside Titusville, Freddie was wide-eyed and hanging on for dear life as we sped along a four-lane through a landscape so unfamiliar to him that for all he knew, we could have been on Q Street in Omaha. We passed a blur of riverside hotels and apartments, filling stations, bait stores, fast-food joints, and tire and auto-repair shops. Through Sharpes and Cocoa we sped, past McDonald's, Tire Kingdom, Arby's, and other franchises, with the breaks between buildings offering blurred movie frames of the river. We passed a car dealership displaying more shiny cars in a single lot than Freddie had seen in a lifetime.

Farther out of Titusville, the development momentarily broke, and we passed by banks of the river lined with slash pine and cabbage palms presiding above a thick understory of foliage. "No fire around here in a while," Paul Schmalzer quipped.

He said that the scrub-jay had declined almost in direct proportion to the loss of the scrub itself, which, of course, was related to human population growth. Between about 1940 and 1990, there had been a 70 percent reduction in scrub. During the same time, the scrub-jay had declined by almost 90 percent.

In Freddie's day, fewer than 5,000 people lived in the county. By 1950, there were still only about 25,000. With the arrival of the National Aeronautics and Space Administration and new air force facilities in the early 1960s, the population ballooned to nearly 100,000. In its heyday, NASA alone employed some 26,000 people, or five times the total county population in 1925. By 1970, the population exceeded 230,000. By the time Schmalzer and I were driving the highway, more than a half million people lived in Brevard County.

It was the space race that drew so many people to the county so quickly. In an unintended consequence, the land on Cape Canaveral that was set aside and protected for the growing space center would one day hold the only viable populations of scrub-jays in the county. But in Freddie's time, the moon was just a mysterious orb that poured night-silver into the Indian River.

About thirty minutes south of Titusville, my cell phone, channeling Google Maps, commanded us to turn right onto Barnes Avenue and then confirmed, "Your destination is a half mile on your right." Already in a state of suspended belief, Freddie could hardly believe what he'd just heard.

The 140-acre Helen and Allan Cruickshank Sanctuary is wedged among developments, with the surrounding houses visible before we even got out of the car. Schmalzer pulled a folded sanctuary brochure from his back pocket—he'd picked it up on one of his many earlier visits—and unfolded it against a wooden signpost. An aerial photograph of the region showed a small dark-green square in the center of a motherboard with parallel circuits symmetrically winding through memory chips and other components. "Another island of scrub in a sea of development," Schmalzer said with a long exhale.

After we walked through an area of sand pine, the land rose slightly, and we entered the scrub. In the distance was a line of backlit sand pines, their tippy-tops imploding in the sun's blazing glare. We stood silently for a moment, hoping to spot a scrub-jay. We pished. The breeze whispered a noncommittal reply.

Then something moved on the ground, as if a patch of sand had begun to levitate. "A gopher tortoise," Schmalzer said. Scrub-jays or not, the scrub at Cruickshank was the best we had seen that day. Burned about a decade earlier, the oaks ranged from waist to shoulder height, with lots of sand openings. The pines in the immediate area were few—enough to serve as sentry posts but not enough to give cover to hawks. I pished again. "We're getting into the heat of the day, so I'm not sure," Schmalzer said.

We came across a tall spike of the cream-colored blooms of the bigflower pawpaw, a member of the custard apple family. Past their prime, the blossoms drooped but clung stubbornly to the chin-high stalks. Its four concave petals, reminiscent of mountain dogwood, have earned the bigflower pawpaw the nickname "Florida dogwood." The leaves of the pawpaw are tough and waxy, able to withstand the heat and drought of the scrub. In a few months, fruits would hang where the blossoms now faltered. The fruit has a soft custard texture and an indescribable flavor, or, as one botanist put it, "isn't the same as anything I have eaten." Schmalzer said the plant is a host for the spectacular zebra swallowtail butterfly.

I was suddenly startled by a thump on the top of my hat. "Well, there's one!" Schmalzer said, pointing to the scrub-jay that had landed on my head. "You got a visitor."

A second scrub-jay suddenly showed up in a nearby bush. As if bringing up the rear, a third bird assumed sentinel duty in a tall dead pine close by. In direct sunlight, the scrub-jays' plumage seemed radiant, a momentary glimmer of aquamarine. The bright blues of their feathers arose not from pigmentation, as in many other kinds of birds, but from a light-scattering spongy layer of keratin in the barbs. Although not technically iridescent, in the right sunlight its feathers seem to sparkle with a blue as intense as the summer Florida sky.

Schmalzer and I watched the scrub-jays for a while, and they watched us. Finding no offerings of peanuts, two of the scrub-jays took flight and disappeared into the low oaks. The sentinel lingered for a while, and then he left too. Observers often tout scrub-jays for their friendliness. Adaptable and opportunistic might be more accurate. Overt friendliness is a sign that visitors often feed the birds.

Even from this brief encounter, it was easy to see why the Florida scrub-jay is so popular with birders—when they find one. A master self-promoter, a scrub-jay is its own best advocate. A seeming model of morality, its lifelong devotion to its mate invites people to project their own aspirations upon it. Its living in family groups in which older siblings help to protect and raise the youngsters—sometimes misconstrued as altruism—earns it a lot of admiration. Its rarity gives it high spotting value among birders. With a brain the size of a chicken's, the bird has a memory like an elephant's and can recall the exact location of thousands of stashed acorns. There can be no doubt it's intelligent. A homebody among birds, the scrub-jay rarely ventures more than a mile or two from its birthplace. Dedicated to mate and offspring, devoted to its territory, and always on the lookout for an easy meal, the motto on the scrub-jay's family crest might read "Family, County, and Peanuts."

Not everyone credits the bird with noble qualities. Several attempts have been made to have the Florida legislature name the Florida scrub-jay as the state bird instead of the northern mockingbird, already so designated by Arkansas, Texas, Tennessee, and Mississippi. The mockingbird is found across the lower forty-eight, throughout Mexico and the Caribbean, and in parts of Canada. The Florida scrub-jay, of course, is Florida's only bird that lives nowhere else, and it's in the exclusive company of just over fifteen or so birds limited to the United States.

I told Schmalzer I'd read a piece by Representative Mark Pafford of West Palm Beach, who had introduced a bill into the Florida legislature to replace the mockingbird with the scrub-jay as the state bird. In an editorial for the *Tallahassee Democrat,* Pafford had accused the reigning mockingbird of usurping the scrub-jay's rightful claim to the throne. Marion P. Hammer, a well-known mockingbird supporter, gun advocate, and National Rifle Association lobbyist, shot back. "Unlike the mockingbird, the Scrub-jay can't even sing—it can only squawk," she wrote. She added that "the Scrub-jay is not an impressive looking bird. It is a frumpy looking little blue and brownish bird that lives exclusively in a very small area of Florida." And the mockingbird "doesn't need government protection or our tax dollars to survive." On another occasion, Hammer criticized the scrub-jay's willingness to accept peanuts from visitors: "Begging for food isn't sweet. It's lazy and it's a welfare mentality," she wrote with apparent sincerity.

A few days after publication of Hammer's opinion piece, John W. Fitzpatrick and Reed Bowman, the world's leading authorities on the Florida scrub-jay, wrote a rebuttal. Scrub-jays "in close contact with one another . . . offer long and quite beautiful series of sweet warbling notes," they replied. As for the scrub-jay being unimpressive, "while nobody has ever visited Florida to see the gray, perpetually nervous Northern Mockingbird, literally thousands of nature lovers from around the world visit Florida each year just to see our unique scrub-jay, and they invariably proclaim how beautiful it is, with striking iridescent blues offset by a sand-colored back and frosty-white crown."

The latest legislative attempt failed. Advocates had hoped making the scrub-jay the state bird would raise its status and increase political support for it. That's what others feared. The scrub-jay was pretty much powerless in the face of development, and a lot of opponents of the legislation wanted to keep it that way.

Schmalzer said he was familiar with the state bird debate but hadn't followed it closely. He'd rather spend his time and efforts in the field. As he cupped his palm above his eyes as a visor and scanned the horizon of the Cruickshank Sanctuary for more scrub-jays, he said, "Well, I think I said we might see a few. We saw a few."

The day was getting away from us. He said there was an even bigger patch of scrub about twenty-five miles farther south in the county at the Malabar Scrub Sanctuary, just on the other side of Melbourne. And so again we launched the car from the scrub island and headed off through a roiling sea of development.

* * *

The traffic along US 1 increased as we drove south. As we waited for traffic jams to clear and red lights to change, Schmalzer spoke about the largely failed efforts to conserve the scrub-jay in Brevard County.

"We had no expectation that any small patch of scrub around the county was going to be a permanent home for the birds," Schmalzer said. "But we did hope that these various islands of protected scrub would form an interconnected archipelago island chain capable of permanently supporting many scrub-jay families. It didn't turn out that way. Instead of stepping-stones of restored scrub among the islands, housing developments were built. The rising seas of development had inundated just about all the potential habitat between."

While there's still hope for achieving this, he said the county's two dozen or so scrub preserves are "too far apart to reliably link one group of jays to the next, although it's still possible for scrub-jays to move between several scrub reserves in south Brevard. Even if other pieces came up for sale, county commissioners would be unlikely to come up with the needed money. So take a close look at the last remnants of the Brevard scrub ecosystem."

As we sped down toward Malabar on what had become a six-lane highway, we cruised past the usual franchises, like Dunkin' Donuts and Burger King. Strip malls flew past the car windows like dashes in a sentence, quite a few built atop the scrub that Freddie had driven through a few generations earlier.

After passing through Melbourne, we crossed Turkey Creek. It was there, nearly two hundred years ago, that Charles Vignoles, a surveyor visiting from St. Augustine, first noted the proximity of pine, scrub, and sandy soil. Perhaps without knowing it, he stated what ecologists would later understand to be among the defining characteristics of scrub—sand dunes, low-growing oaks, and a smattering of pine. "Immediately beyond the mouth of this creek are the Turkey Creek bluffs, of rich yellow sand and forty feet in height, extending a mile in length; a trifling distance further on the bluff is of shells [sic] with a scrub hammock, the northern sand bluff being covered with pines, and having a luxuriant under brush of oak and hickory scrub," he wrote. When Freddie drove through the area in the 1920s, much of the scrub Vignoles described still existed. In its place now are Anglers Drive Northeast, Herndon Circle, Ridge Road Northeast, and Riverside Thai and Sushi Bar. It has probably been a half century or more since scrub-jays lived there.

Soon the six lanes collapsed into four. On we went through the post-Freddie world of Brevard County, past weedy and scattered open spaces, a Deals on Wheels used-tire shop, Citgo, and all the scattered artifacts of free enterprise poured without a plan in concrete onto a disheveled landscape that throughout Freddie's day had been home to thousands of scrub-jays. Today, along the entire length of US 1, there's as much chance of glimpsing a scrub-jay along the highway as getting to Atlantis on a moonbeam.

I was beginning to feel as if we were two honeybees on a desperate, final harvest before fall, traveling from flower to distant flower, all of them wilted. Finding little nectar, we alighted and flew on and on, straight into winter.

Soon we came to our final stop—the four-hundred-acre Malabar Scrub Sanctuary, less than a half mile from the Indian River and two miles south of Palm Bay. Just inside the entrance gate, we came upon what appeared to be an abandoned four-lane highway with a median overrun with palmettos and bushes. It was an eerie sight, like a city mysteriously abandoned. Schmalzer said we were now standing in the grand boulevard through what was to have been a development—before the crises of the late 1980s and early 1990s, a time when one-third of the country's savings and loan associations failed, taking with them the dreams of numerous developers throughout Florida. The road even had a name: Malabar Woods Boulevard. The builder had gone bankrupt before houses were built. Shortly after, the county bought the property on the cheap.

Most of it was scrub. Two parallel ridges of scrub-jay habitat ran through the sanctuary. A vein of pinelands and some hardwoods filled the lower land between them. We walked for a while but saw no scrub-jays. Only a few live there now, Schmalzer said. Still, there is hope that scrub-jays can persist at the site as more restoration occurs. Such a hope has all but gone throughout mainland Brevard, but perhaps there is something to be said for hoping for hope's sake.

At Malabar, I was again struck by the sere western feel of the chaparral-like scrub where we stood. Schmalzer and I talked about the western origin of the scrub-jay and some of the plants around us. But the west traveled both ways. It was in the scrub and other areas of Florida that "western" cowboy culture was born. Long ago, the original vaqueros in Florida and their descendants developed the tough, small breed known as Florida scrub, or cracker, cow. And it was the vaqueros from Florida who introduced cowboy culture and its iconic accoutrements like the lasso west of the Mississippi. If the classic western television series *High Chaparral* had been filmed in the Malabar Preserve, no one would ever have known the difference.

Schmalzer and I had many conversations as we traipsed through the remnant scrub of Brevard County, and I knew I would miss them. Before heading back to Titusville, we stopped for late lunch at the Shack Seafood Restaurant on the Dixie Highway in Palm Bay. From the dining room, we had a sweeping view of the Indian River. We had arrived at the tail end of a major algal bloom in the lagoon caused by an overflow of polluted water from Lake Okeechobee, about seventy miles south and inland. Where the Indian River's surface usually fluttered with white sails or the wakes of powerboats, it was mostly deserted because of the smell and eye-stinging fumes. Even the waterbirds had mostly fled.

I mentioned to Schmalzer how surprised I was that scrub-jays had been so hard for us to find. "Come back in a few years and it will be a lot harder," he said. He said that except at the Merritt Island National Wildlife Refuge at the Kennedy Space Center and on air force land on nearby Cape Canaveral, the last opportunities for saving self-sustaining populations of scrub-jays are quickly dwindling. With enough money, restoration, and management, scrub-jays hanging on in several larger conservation areas in the northern part of the county might survive. Sadly, history says they will not.

The population at the Cruickshank Sanctuary we visited earlier in the day and adjacent private conservation land offer glimmers of optimism. With enough restoration and management, the conservation lands of the south end of the county, such as the Malabar Scrub Sanctuary, might one day harbor a permanent population. St. Sebastian River Preserve State Park, a short walk from Freddie's former home in

Sebastian, encompasses about twenty-three thousand acres, including several thousand acres of scrub and several families of scrub-jays.

There is also some hope for a viable population along the Atlantic Coastal Ridge at Jonathan Dickinson State Park in Martin County, sixty miles south of Sebastian. In the early 1990s, the park was home to nearly three hundred of the birds, but by 2012 the numbers had crashed. Now, with restoration under way and fire on the landscape once again, the numbers are rebounding. Today there are nineteen scrub-jay families at the park, up from ten just a few years ago. The scrub within the 10,500 acres could hold up to eighty families, though the park would have to more than double its current population to become a "source," or a place where the population is growing. Still, it is the best hope for scrub-jays to persist along the southern Atlantic Coastal Ridge on the mainland.

Having expressed some optimism, Schmalzer shook his head and added, "At one time, the coast from St. Sebastian to Jonathan Dickinson had one of the highest concentrations of scrub-jays on the Florida Atlantic Ridge. Almost all of this has now been developed and is either golf courses, malls, highways, or wall-to-wall housing. Look at it on a map. It's unbelievable." He said if I wanted to see scrub-jays in Brevard County, I would have to get off the mainland and visit Merritt Island and the Kennedy Space Center.

After dropping Schmalzer off at his Titusville home in a grove of pines, I was still hungry and drove into town and had a couple of scrambled dinosaur eggs at the local IHOP. Later, I drove to a spot near Space View Park along the Indian River across from Merritt Island. I took the picture of Freddie off the dashboard and held it so he could see across the water and, on the horizon, the Vehicle Assembly Building at the Kennedy Space Center. "What the hell is that?" he asked. I told him they built rockets there. He shook his head and said I was crazy.

4

KENNEDY SPACE CENTER

Not long after my trip with Paul Schmalzer, I took his advice and arranged a visit to the Kennedy Space Center and the Merritt Island National Wildlife Refuge, the last stronghold of the scrub-jay on the Atlantic coast. The wildlife refuge encompasses nonoperational areas of the facility. Although the Kennedy Space Center has had a lower profile during the decade since the National Aeronautics and Space Administration ended its Space Shuttle Program, a lot was happening around the time of my visit in 2016. Wealthy entrepreneurs such as Tesla's Elon Musk and Amazon's Jeff Bezos were leasing NASA land for their private space companies SpaceX and Blue Origin.

A few months before my visit, technicians for Musk's SpaceX were filling a container inside a liquid oxygen tank at Space Launch Complex 40, when the Falcon 9 rocket blew up, taking with it a $195 million communications satellite owned by Facebook and sending the scrub-jays living nearby diving for cover. The mishap set the California-based SpaceX back tens of millions of dollars and months behind in its plans to take astronauts into space and, eventually, to colonize Mars. Regarding his goal of Mars colonization, Musk had earlier opined that there is a "strong humanitarian argument for making life multi-planetary" and that people have a duty to "safeguard the existence of humanity in the event that something catastrophic were to happen."

But something catastrophic already is. If the arrival of the Anthropocene epoch, with tens of thousands of species being pushed to extinction and melting ice caps threatening coastal cities and the health and welfare of tens of millions of mostly poor people, isn't catastrophic, what might be?

The whole space enterprise began fifty years earlier with a some-odd-trillion-dollar lunar journey, from which astronauts returned with an eight-hundred-pound treasure of rocks. Now, humans were gearing up to send people thirty-five million miles to Mars at the cost of at least $6 billion for the trip itself and tens of billions more on research and development. During this past half century of space exploration, propelled largely by the dream of discovering life on other planets, we've hastened the extinction of thousands of species on Earth. Seeing the astronomical numbers spent on space exploration, worthy as it is, practically blows the eyeballs out of weary conservationists desperately scrounging for a few million federal dollars to explore and protect life on Earth. The greatest breakthrough in the history of human exploration will not be finding life on a distant planet but discovering the special value of species on Earth.

The scrub surrounding Launch Complex 40 is a fragment of one of the rarest ecosystems in the world. The World Wildlife Fund classifies scrub as critically endangered globally and considers the remaining thirty or forty thousand acres remaining in all of Florida an ecoregion unto itself. Despite its naturally limited distribution, Florida scrub also contains a disproportionate share of endemic species within the North American Coastal Plain, a region Conservation International recently declared to be a global biodiversity hot spot, like the tropical savannas of Brazil and the forests of Madagascar.

Remarkably, a few scrub-jays continue to live in the deteriorated scrub around Launch Complex 40. Fortunately, there's a lot more scrub at the Kennedy Space Center—tens of thousands of acres. The space center holds the last area of scrub in Brevard County big enough to support a stable population of scrub-jays, and it's one of only three "source" areas for scrub-jays in the state. That's a tiny fraction of a scrub nature created to begin with. Naturally rare since it occurred mostly on remnant sand dunes from eons past, now it's approaching the vanishing point.

On the day of my visit, I arrived at the refuge headquarters on Merritt Island around noon. There I was met by Mike Legare, a supervisory biologist with the US Fish and Wildlife Service, the federal department that oversees it. As we sat down at the wooden table in the conference room at the refuge, Legare said that NASA had come by all this scrub incidentally when it began buying up more than 100,000 acres to build the Kennedy Space Center in the late 1950s.

A good 60,000 acres of that land was dry upland, a major reason why NASA bought it in the first place. Originally, up to 80 percent of these uplands were scrub. Today, about 17,000 acres of prime scrub-jay habitat remains. Legare said NASA uses only about 6,000 acres of the space center for operations and has placed the remaining 134,000 acres under US Fish and Wildlife Service management. The Merritt Island National Wildlife Refuge was established in 1963. The national wildlife refuge system actually grew from the same 1903 conservation program launched with the establishment of Pelican Island off the riverbank in Sebastian near Freddie's home.

A few minutes after I got there, Dave Breininger, a scrub-jay biologist who, like Schmalzer, works for NASA and who has studied the scrub-jays at Merritt Island for almost forty years, joined the conversation. He has a PhD in conservation biology from the University of Central Florida and has published numerous papers on the scrub-jays at the space center and in Brevard County. More than encyclopedic knowledge, his logic is unforgiving. But he also has a lot of feelings about what's happening to the scrub-jay. Breininger said that the seven hundred or so scrub-jays that make up these families constitute the third-largest population in the state—surpassed only by the population in the Ocala National Forest to the north and on the Lake Wales Ridge in the central peninsula. While the refuge is among the few remaining "source" areas for scrub-jays—places where the population is growing—just about everywhere else, like the place Schmalzer and I visited on the mainland, are "sinks," which is to say, places where the birds are going down the drain. The biggest single source at the refuge was an enormous scrub known as Tel 4, named for a telemetry station at the site used to track rockets.

"We'll drive over there shortly," Legare said. "Scrub-jays have a bright future here. Scrub-jay management is our number-one refuge priority. We have around three hundred scrub-jay families." He said the responsibility for them was "awesome."

When asked if the population was growing, Breininger lifted his right hand above his head to represent the high point of an imaginary graph and said, "This is where the population used to be." Then he drew a steep curve downward with his hand to about table level. "This is where they are now. Very decreased but leveled off." He attributed the leveling off to the restoration of habitat at the space center. Without those ongoing efforts, the decline would have continued.

"Stable?" I asked.

"Let's just say 'level.'"

"Sure, NASA has helped to preserve the scrub, but there's still a lot less here now than when NASA first bought the land," Legare said. "It's also true there would be very little left if NASA hadn't bought it. But my point is there could be a lot more scrub and scrub-jays if the land was burned more often and then managed for scrub-jays."

Breininger explained that while NASA was, in fact, managing land for the birds, it "managed for the minimum" rather than for the largest potential population. In other words, management was geared toward keeping the bare minimum number of birds required to prevent further decline rather than toward increasing the population. "If all the available scrubland here was regularly burned and managed, twice as many scrub-jays would live on the refuge," he said in frustration.

Legare then suggested we drive out to Launch Complex 40 and over to the expansive scrub at Tel 4. We left the refuge headquarters in a white US Fish and Wildlife truck and, after being cleared at the security gate, drove past the shuttle runway and through an undeveloped stretch known as Happy Creek. The four-mile drive took us past the "space industrial complex," the center of the NASA hive, which included the gigantic Vehicle Assembly Building, or VAB. We drove past the enormous "crawler," a basketball-court-size platform with massive steel tracks that transported the upright shuttle from the VAB to Launch Complex 39. As we turned right on Cape Road, the gantries of Launch Complexes 41 and 40 came into view. "That's where the SpaceX explosion was," Legare said, pointing over a wall of overgrown scrub to the charred remains of Launch Complex 40.

As we drove by it, Breininger shouted, "There's one!" I was surprised by his excitement. Surely he'd seen thousands of scrub-jays over his career. Yet it was as if this were his first. To most biologists, every sighting of a scrub-jay is new in its own way—the joy of serendipity, rediscovery, a glimpse of hope in a bird so rare. Legare quickly stopped the truck on the side of the road, and we got out.

As we walked closer, the bird seemed to rise vertically on the breeze. Then it landed on an oak bush before hopping to the ground, warily approaching us. Surely this is the behavior of a scrub-jay expecting a peanut.

As the scrub-jay eyed us, it hopped through the grass between the road and the edge of the scrub. As its feathers caught light streaming through openings in the thick wall of foliage behind it, they shone turquoise.

Legare spotted a second scrub-jay posted in sentinel duty on the bare branches of a taller shrub.

"The mate?" I asked Breininger.

"For sure."

It remained perched atop the oak bush, glancing at its one-and-only on the ground and watching us as we approached. A third bird suddenly flew up from the bushes. Legare tossed a peanut, and the bird quickly absconded with it.

"It's the whole family!" Breininger said with great delight.

He said their order of appearance, seemingly haphazard, was well choreographed. First came the solitary bird—the dominant breeder male. Soon his mate took up sentinel duty. After a few moments, a second helper ventured out. They watched us for a long time. Then, correctly surmising we were out of peanuts, the family disappeared back into the scrub surrounding the charred remains of the launch complex.

As we got back in the truck and drove toward Tel 4, Legare said that despite the vast amount of scrub at the space center, most of it was overgrown and unsuitable for scrub-jays. Because burning is so limited, the seventeen thousand acres of optimal scrub aren't increasing.

When I asked Legare why they didn't burn more, he looked at me as if to say that a long answer was on the way. "The strict protocol restricts the timing and extent of the fires. We have to burn whenever we can. Or whenever NASA lets us." He said NASA doesn't like fires because the particulates in smoke can drift for miles and blemish telescopic mirrors and other sensitive satellite and rocketry components.

"The NASA people are not concerned with us burning something down. They're concerned with smoke getting into either a facility or a piece of equipment. You could ruin some of the sensitive equipment. When there are very sensitive parts here, they don't want anybody spraying paint. They don't want anybody spraying herbicides or insecticides. No aerosols in the air because they can carry a long way in the wind and damage components of, for instance, a mirror in a space telescope. They don't want anything volatilized."

The answer would have been much longer had Legare not made a quick detour onto a side road to a construction site next to an abandoned orange grove. "This is where Jeff Bezos's rocket manufacturing is going to be," he said as we pulled over next to the large steel frame of a structure that would be the center of Blue Origin. Pointing back to the grove, Legare said, "NASA is increasingly leasing land and supporting services to private space companies. We're trying to encourage NASA management to use land like this grove, where the scrub is already destroyed, for leasing to private space ventures. We don't want the entrepreneurs going into the good scrub. Everyone's heard of SpaceX and Blue Origin, and there may be more billionaires interested in locating their space operations here. We're worried that there could be a time when NASA will start getting squeezed for space and begin to go into the scrub. We hope not."

After the detour, we got back on Kennedy Boulevard and headed toward Tel 4. We turned onto a sand road where an elevated white geodesic dome stood like a big mushroom growing from the scrub. That was Tel 4. Sitting beneath the dome were two trailers resembling the kind of containers eighteen-wheelers pull. A pair of red and blue lights intermittently flashed at one corner of a container. "You'll notice it's an unmarked facility. A lot a weird stuff going inside. It's for tracking 'space junk,'" Legare said, with air quotes.

"And a whole lot more," Breininger added. "Nobody knows what goes on in there."

Legare continued, saying more than he probably meant to. "When people have an animal problem, when a snake or alligator or bird or other critter gets inside a building, I'm the guy they call, and I go in to take care of it. I've been everywhere at this place and inside a lot of places, including the top-secret ones. But I can't tell you what goes on inside *that* building because I don't even know."

The truck slowed as the sand ruts deepened. Breininger pointed out the window to the area known as Tel 4. Low shrubs covered it from horizon to horizon. I stuck my head out. It was by far the biggest expanse of Florida chaparral I'd yet seen.

5

TEL 4

We got out and stepped across a narrow, wet roadside ditch and up a two-foot embankment onto dry sand. "We're climbing thousands of millimeters," Legare said, borrowing one of Breininger's favorite phrases for describing how a few feet of change in elevation can spell the difference between wet and dry habitat in Florida—or between pineland without scrub-jays and a scrub teeming with them. "We're gaining altitude," Breininger added. "Having any trouble breathing yet?"

On aerial photos I had seen hanging on the wall back at the refuge headquarters, the Tel 4 scrub consisted of curious parallel ridge-and-swale formations—alternating parallel lines of sand-covered ridges running alongside greener gullies. Tel 4 looked nothing like that from the ground. At eye level, the dominating scrub completely masked this complex topography. The concealed lay of the land dictated exactly how the scrub-jays were distributed across it. They stuck to the scrub-and-sand-covered ridges.

We walked farther into the scrub.

"Earlier in the day the birds were all over here, everywhere," Breininger said, sounding a little disappointed. "They went into the bushes as it got hotter. Either that, or maybe there was a hawk out." Legare added, "You can hardly imagine how the birds are all over here when they're out."

Still, I could imagine them undercover, each family within its territory, crouching low in the shadow of the thicket and cautiously tilting an eye skyward through the tangle of branches in search of their hook-beaked nemesis. Or perhaps napping, sending out a beep or two to check on the location of the others. Living in a now-you-see-us-now-you-don't habitat can give the impression that birds are scarce. But knowing of the scrub-jays' well-documented abundance at Tel 4—about one hundred families—was good enough—especially compared to the smattering of jays scattered across the entirety of mainland Brevard County.

As we stood in an open patch of sand amid that thick growth, Legare and Breininger spoke in an alternating chorus of optimism and despair. For every reason to find hope, they found two reasons to be discouraged. While the scrub-jays there are holding their own, the number is far below what they could be with more burning. The fewer the birds, the greater the probability of extinction. Yet, there was such promise here.

Satisfied that regular controlled burning had turned Tel 4 into more than 2,500 acres of prime scrub-jay habitat, Legare also acknowledged that truly natural scrub, with its various stages of growth and large patches of open sand that scrub-jays need, would probably never return without natural fires. And if NASA had anything to say about it, before that happened the moon would grow palm trees.

"Scrub-jays need a big territory, about twenty-five acres per family, and they don't like too much tall stuff. They need a height like this that's ten years old," Legare said, touching his shoulder with his palm. "They need some of these open sand patches for scavenging and caching acorns. If scrub isn't burned after about ten years, openings like these will start to close. They're very hard to get back. They're a mystery. We still don't know what causes them. We work collaboratively trying to come up with ideas about how to get the bloody openings back. None of this stuff came down to us in a book about how to do it. As plant ecologists and managers and animal ecologists, we all find ourselves scratching our head sometimes about how the scrub works."

"Now we have all those nice polygons on the refuge," Breininger lamented. "Sometimes we end up with patches of scrub that might be a thousand acres and that don't look all that much different from a cornfield because all the vegetation is the same height. They have the right plants, but the arrangement is a bit peculiar."

Breininger continued to talk about "managing for the minimum." "With better restoration, the current population could easily be almost doubled," he repeated. "A politically acceptable number of birds differs from a biologically acceptable number. When you manage a species for a minimum number, you take away that buffer and increase the probability a population could be wiped out by a catastrophic event like a hurricane or disease outbreak. If you want to keep a species around for a long time, you have to manage for the maximum." And the only way to manage for the maximum, both he and Legare agreed, was to burn, burn, burn.

As if NASA's limitations on burns weren't enough, human changes to the landscape make it harder for fires to spread once they are set. "Airports and other facilities—on and off the space center—create firebreaks and can even shift the direction of winds, both of which can suppress fire," Breininger said. "Changes in the landscape have altered the whole pattern of fires here and all over Florida. At the space center, if a patch of scrub goes, say, twenty years without fire, you often can't burn it without first cutting down trees and removing some of the heavy brush to reduce the fuel load. Otherwise, the fire burns too hot and kills trees and scrub plants it's not supposed to."

As we spoke, the sun rose higher into the December morning. With the increasing onshore breeze, the faint scent of Atlantic salt spray reached us from half a mile away. Five miles north, the VAB was visible on the bright horizon and, far to the right of that, Launch Complex 40.

"Sometimes it's money, sometimes it's operational issues with NASA, sometimes it's our staff, sometimes it's not having the right training or enough people for burning. Sometimes it's just that fires won't start," Legare continued. "Fires we set don't necessarily catch and burn anything. Some quickly burn out, leaving only a tiny patch of smoking ground. We would technically say it's been 'treated,' but it's not really a burn at all. Of the fifteen thousand acres we 'treat' annually, only about ten thousand catch fire and burn. We have so many rules. As a start, every one of the 'units' of land on the refuge has a legal document that sets what conditions it can be burned under. They have to be reviewed by independent fire experts. Then it can be approved by NASA or not. The paperwork accumulates. Every year it gets more complicated.

"In the prescribed burn business, you don't just pour gasoline, light a match and run, and come back the next day hoping the job is done," Legare said. "You need a certified 'burn boss,' qualified by the National Wildfire Coordinating Group, who lights the fire. Then there's the 'ignition boss,' who oversees the whole procedure. There are different types of burns and different categories of moisture, such as 1H moisture, 10H moisture, 100H moisture, live herbaceous moisture, live woody moisture. Weather, midflame wind speed, upslope, slope steepness. All go into a computerized burning model, which, in the best of all worlds, will predict fire behavior on that day.

"You also have to have the land unit prepared so that the firebreaks around it are maintained and we can physically get it done. Then we need to have the people. We need to have the money, of course. Then we need to have permission at the same time as we have the appropriate weather."

Legare was growing exasperated, and he suggested we drive over to the beach. We passed the VAB and the industrial complex once more, then drove along Saturn Causeway, past Launch Complex 39A, onto Cape Road, and finally onto a narrow, sandy byway right into the scrub and just a few yards from the beachhead. The change from interior scrub to coastal scrub was obvious. Both are considered scrub, and scrub-jays inhabit both. Of these two general classifications, the coastal variety is far rarer. Although once dotting hundreds of miles of white sand and certain coastal soils from the Atlantic Ocean around to the Gulf of Mexico, the surviving areas can now be measured in square yards rather than acres.

We were entering one of the largest remnants of coastal scrub in the state. The skinny two- or three-thousand-foot strip had already been sliced in half the long way by Cape Road and a parallel access road between launch complexes. This left only a severed contiguous strip about six hundred or seven hundred feet wide hugging the shore. But there it was, hanging on despite all the historical disruptions—and now the rising sea level was getting ready to deliver the coup de grâce.

This coastal scrub, also known as coastal strand, was more open and airier than the denser interior scrub at Tel 4. Coastal scrub is determined not only by its nearness to the shore but also by peculiarities of the soil that support it. Still dominated by scrubby oaks, coastal scrub's community of other plants differs from interior scrub. The coastal scrub was an awesome garden, beautiful and distinctive, often dominated by saw palmetto, sand live oak, myrtle oak, and various mixtures of yaupon, railroad vine, bay bean, sea oats, sea purslane, sea grape, Spanish bayonet, and prickly pear. It brought back memories of the small home we lived in on the beach at Indialantic when I was a child and the gardens of sea oaks, sea grapes, and the scrub along the beachhead. Legare, Breininger, and I lingered for awhile, taking in the surf, sounds, and smells and the vanishing habitat of coastal strand. They said scrub-jays could often be found here. But not today.

We got back in the truck and drove a ways farther along the coast road and stopped near a dune just behind Launch Complex 39A, home of Apollo and space shuttle launches. "Now I want to show you something—why NASA has so much on its mind these days," Legare said as we got out and walked over to a berm. Legare pointed out old railroad tracks used by NASA in the 1970s and 1980s. They were rusty, and many of the ties were displaced, and the sand had been washed out between many of those that remained. The rail line was on the verge of being swept from the dune and into the sea.

"This used to be quite a bit inland," he said. "It shows you how far the sea has risen." A few hundred yards to the north, NASA had built a protective dune between the beach and the launchpad. "Only it mostly got washed away by the recent hurricane," he said.

The potential impact of climate change on the scrub-jay has barely been studied, Breininger said, but the risks were increasing. The more frequent flooding of scrub such as that in Tel 4 because of higher sea level and larger, more frequent hurricanes was the most obvious. The increasing temperature could change the timing and production of the acorns the scrub-jays rely on, while more intense drought could further reduce the frequency with which controlled burns were allowed.

According to a report by the US Global Change Research Program, by 2100 the Kennedy Space Center could have up to forty more days of plus ninety-five degrees Fahrenheit every year. The higher temperatures could force scrub-jays to seek new refugia in cooler microclimates provided by streams with cold-water upwellings.

"So many problems," Legare said as we climbed back into the truck. He sat for a moment as if to compose himself. Then he turned the ignition, and we rode back to the refuge headquarters mostly in silence.

PART TWO

THE GULF COAST

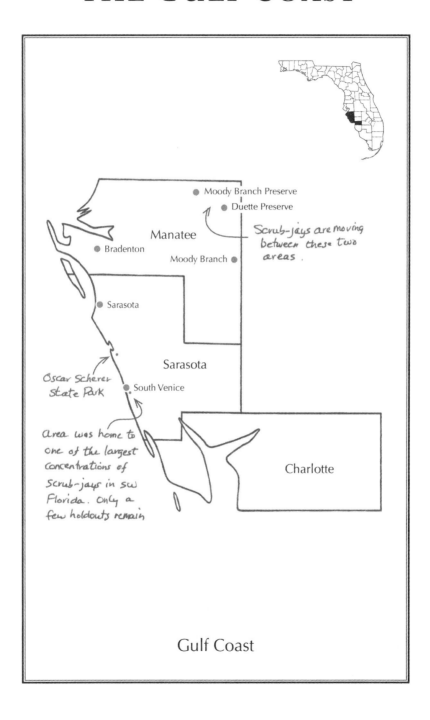

Moody Branch Preserve

Duette Preserve

Manatee

Scrub-jays are moving between these two areas.

Bradenton

Moody Branch

Sarasota

Sarasota

Oscar Scherer State Park

South Venice

Area was home to one of the largest concentrations of Scrub-jays in sw Florida. Only a few holdouts remain

Charlotte

Gulf Coast

6

SARASOTA COUNTY

If a birder in the mid-1990s had traveled from California to the Sunshine State to see a Florida scrub-jay—and many birders did just that—he would have bragged to the folks back home how rare the bird was. After all, he had to travel thousands of miles just to find one, and then, even after reaching Florida, he had to find the right spot—say, the scrub-jay stronghold of Highlands County. Even a visitor from Tallahassee to Highlands County might have commented to people back home of the bird's rarity. But if you'd said to a Highlands County resident, "That bird sure is rare," he'd have looked at you incredulously and replied, "They're not rare. I see them around here every day!"

The Florida scrub-jay represents a population paradox. On the one hand, it's rare, but on the other hand, it's common (or at least once was). It kind of depends on where you're from. For that matter, you could say the same thing about surf shops, which are rare to someone living in Topeka but common if you live in Titusville. Biologists have the perfect term to resolve this paradox. They call species such as the Florida scrub-jay "rare but locally common."

To most laypeople, a bird either is rare or not, but biologists define different types of rarity. One way to look at it is whether a species has a large or small population. Another is whether the population has a large or small geographic range. Finally, there is the question of whether the species, regardless of the first two characteristics, exists across many types of habitats or is restricted to a narrow one. By categorizing a species on all three accounts, biologists define different kinds of rarity and can get a better read on the vulnerability of that species to extinction.

A "common" species scores in all three categories: it has a large population, has a broad geographic distribution, and is distributed across a broad range of different habitats. If it doesn't meet all three of these criteria, then it's "rare." But what kind of rare? For example, a species can be rare because it has a large population in a limited area. Or it can have a large population across a wide geographic area but be limited to a narrow range of habitats within that larger area. If you do the math, there are seven unique combinations of the three factors. One of these combinations fits the scrub-jay. It has a limited geographic range, is restricted to the scrub habitat, and is abundant in at least a few areas. Therefore, it's said to be "rare but locally common."

I mention this population paradox because it's easy to confuse local abundance with overall rarity and from there fall for a logical fallacy: the scrub-jay is by nature rare, and it's done fine being rare for two million years; therefore, even though it's still rare, it will do fine for the next two million years.

Why worry?

Because the nature of its rarity has changed. It is no longer common locally. The scrub-jay has always faced the double jeopardy of a naturally limited geographic distribution and restrictive habitat requirements. Now the final bulwark of its survival—local abundance—is giving way. The scrub-jay faces a triple whammy that greatly increases the probability that it will become extinct.

Those who say the bird will be fine will often, when pressed, roll out a different argument. That if the bird does go extinct, then it must have been nature's way. After all, they ask, don't all species eventually become extinct? Yes, but that's totally beside the point. Natural extinctions occur across an immense geologic time span, not as measured by the Gregorian calendar. It would be like a geologist trying to measure geologic change by staring at a rock while glancing at his wristwatch. He's not going to get the data he needs because he's terribly confused about timescales. If he wants to see short-term change, he needs to pick up the rock and smash it against another rock. But that's not a measure of natural change any more than "smashing" a species like the Florida scrub-jay has anything to do with natural extinction.

Curious to document some examples of how a bird that was once locally common had become so rare, I decided to visit Sarasota County, a place where the scope of biological destruction takes on biblical overtones. I'm not sure exactly which sins against the Bible best apply here—there are at least seven—but let's just say that what's happened to the scrub-jays there is exactly opposite of what happened on the fifth day of Creation, when God proclaimed, "Let birds fly above the earth across the dome of the sky."

I decided to start with a foray into the county's recent past at Oscar Scherer State Park, a forlorn 1,300-acre land of hammocks and scrub not far from the city of Sarasota. Beginning in 1955 with a humble 462 acres of former ranchland donated by Oscar Scherer's daughter Elsa Scherer Burrows, the park gained another 900 acres from Palm Ranch in the mid-1980s. Without scrub-jays at the time, the overgrown park could best have been described as the land that ranchers got and fire forgot. But once regular burning was reinstated after incorporation into the park, the birds returned. However, for several unexpected reasons I'll get into—spoiler alert—this story doesn't have a happy ending. Once a beacon of hope for the scrub-jays of Sarasota County, Oscar Scherer became the setting for a cautionary tale, which leads to the forlorn part.

Nobody understands this melancholy saga better than Sandy Cooper, eighty-six, a park volunteer and, for the past seven years, unofficial census taker of the remaining birds there. He knows each bird's leg band colors and name and counts the birds every month. He is the first to record when another one goes missing.

"I kind of look at them as if they're all my friends. They know me. They know I always wear the same green clothes and the same kind of hat," Cooper told me early one morning in the spring of 2016, as we were accompanied by Tony Clements, the park manager.

An avid birder, Cooper had a life list of 655 species in the lower forty-eight states and a list of 2,000 species in thirty-five countries. In his professional life, he worked on the research and development team at Procter & Gamble that developed Crest toothpaste and Head & Shoulders shampoo. He later became the president and chief executive officer of Burke, a major market research company in Cincinnati. But volunteering as census taker for the scrub-jays at Oscar Scherer was, he said, "one of the most important jobs I've ever held."

Cooper said that when he retired to the area from Ohio, his "neighborhood was just being built, new houses going up, a lot of territory still open and eastern bluebirds around, nice birds; bobwhite quail were there. Being a bird-watcher, I'd see them in my backyard, three or four other species. Today they're not there at all. The reason is some birds just don't like being around people and development and houses and cars and things, and when that happens, they move. Scrub-jays are country people. They would rather be in a different place."

In the decade before Cooper arrived in Sarasota County, in 1997, housing developments were going up everywhere, roads were being built, and the scrub-jays left in the county were rapidly declining. The county's two hundred or so survivors had for decades eked out an existence in mostly small, degraded patches of scrub that dotted the surrounding suburbia. They were scrub-island castaways.

This included a few scrub-jays living about a mile and a half southeast of Oscar Scherer in a wedge of scrub between I-75 and the Calusa Lakes Golf Club. Less than a mile from there, a few jays lived on a scrubby patch of unprotected land squeezed by I-75, Laurel Road East, and Kings Way. About five miles southeast of Oscar Scherer, a few dozen birds survived in remnant scrub around Venice Municipal Airport, some of them in a patch 1,500 feet off Runway 31. A few families lived in an archipelago of tiny scrub islands ten miles south of Oscar Scherer at Service Club Park, next to Sharky's on the Pier,

Caspersen Beach, and Shamrock Park. These were the remaining emeralds of what once had been a dazzling necklace of scrub along Sarasota's coast. Knowing that these populations were on their way out, state officials rushed to restore the scrub at Oscar Scherer and give these castaway scrub-jays a real home.

In 1992, volunteers and state employees cut and hauled off the largest trees and brush and then set fire to the remaining stumps and vegetation. Within weeks, lime-green palmetto sprigs appeared on the charred ground. Within a few months, the roots of the burned oaks sent up crimson sprouts, which soon turned bright green. Having seen regular burning for the first time in almost a half century, the scrub—hidden in the roots and seeds beneath the sand—still remembered what it was. And that is what it slowly and surely became again.

Scrub-jays began showing up at the park for the first time in anyone's memory. Like refugees returning home from war, the birds flooded the sprouting scrub at Oscar Scherer. Fleeing the tiny patches of degraded scrub at the edges of housing developments outside the park and from other scattered outposts, the birds came solo, by pairs, and in families.

By 1997, 175 birds lived at Oscar Scherer. By 1999, the park held the largest single population not only in Sarasota County but on the Gulf Coast. Twenty of the birds were hatched inside the park, signifying that it had become home. What seemed like a miracle was just common sense: restore the scrub and the birds will come.

With a few hundred acres of rejuvenated scrub now available in the park, biologists predicted the population would peak (it did so in 1996), fluctuate, and eventually stabilize at about twenty breeding pairs. As the theory went, the population would expand until the park exceeded its carrying capacity, and the scrub-jays without territories would hold out in nearby inferior habitat until more scrub was restored.

But by 1999, three years after the peak of 175, only 150 birds were living in the park. By 2002, only 100 remained. Then there were 60. Birders, biologists, and conservationists were flummoxed. As Cooper witnessed the unexpected decline, he asked himself, How could such a sweet story of success be ending so badly?

One problem was that scrub outside the park had not been restored or expanded as had been planned. Lacking suitable territory outside the park for spillover birds to expand back into, the Oscar Scherer scrub-jays squabbled over space. Frequently skirmishing with aggressive younger birds without territories, the birds in the park apparently had little time for nesting and raising young.

Surrounded by housing developments and domesticated pets, the park was invaded by domesticated cats, which stalked the birds and small mammals. The developments also funneled natural predators such as snakes, raccoons, and bobcats into Oscar Scherer.

Not to mention that the bird feeders in surrounding yards may have upset the birds' diets. And that cars speeding along a nearby highway killed some birds trying to cross. The number of scrub-jays at Oscar Scherer kept falling.

Todd Mecklenborg, a biologist with the US Fish and Wildlife Service and leader of the Florida Scrub-Jay Recovery Team—a group mandated by the Endangered Species Act to develop a strategy for protecting the bird—said, "When that land was acquired for Oscar Scherer, one side had development. Now, four sides are developed. Sarasota County just never took management for scrub-jays seriously, and the Fish and Wildlife Service just kept giving permits to 'take' scrub-jays," he said, referring to a special approval that allows home builders and others to "incidentally" harm scrub-jays. "This forced scrub-jays into smaller and smaller areas. Oscar Scherer became overpopulated, and the birds bottlenecked. When they tried to disperse, nothing was available for them to get to. Other patches of scrub are so embedded in the urban interface that fire management is difficult, and the county commission has made it very much more difficult. If anyone complains, you have to put the fire out. Period. The population at Oscar is ultimately going to become locally extinct."

By 2016, only six scrub-jay families remained in the park, each with its own territory. In May, when I visited the park with Cooper and

Clements, only five were left. "It's a very sad story," Cooper said. "I've had the sad privilege of watching it happen. One by one by one."

During my visit, Cooper, Clements, and I took a ride in the park vehicle into the scrub. As we approached the first territory—it was marked 2A on the map they gave me—Clements stopped on the sandy road. He got out and began pishing the jays—*shreep, shreep, shreep.* None were to be seen. But Cooper knew they were in there somewhere.

"The five birds currently living in 2A are an accidental family," he explained. "One day a bird in a different territory lost its mate, and it came over here and paired up with a lone 2A bird. Their chick fledged about a year ago. It was the only pair in the park to produce a fledgling in 2016. Then two outsiders came in. One came from a nearby territory and the other from outside the park. The outsider arrived without notice one day; didn't tell us where it came from; didn't have a band, so we didn't know. Often when somebody new tries to come in like that, the existing family drives them away. But in this case, both new birds were accepted by the family. The 2A family of five birds is huge for us. That's one-third of everything we've got in the park."

"It's not looking good," Clements added. "The fledgling we had in 2016 isn't just the only fledgling in the park; it was the only one that made it in all of Sarasota County."

As we waited for birds to come out, Clements pointed out a horde of yellow-and-black-striped caterpillars with reddish and black bristles crawling on the ground next to us. "Scrub-jay food," he said. "Every five or six years they will get like this and will be everywhere. The paved road into the park becomes slick with them because as they are run over, more caterpillars come out to eat them and then those are killed, so it's a vicious cycle. We've had it where campers leave because they think it's raining because the caterpillars are pooping from the trees."

Clements said that the abundance of the caterpillars—they're the larval stage of a native moth—peaks with the abundance of lupine, their host plant. "The scrub-jay babies also hatched when those caterpillars were abundant. Mother Nature's got it all set up—lupine, caterpillars, and baby scrub-jays. We don't need to get in the way and mess that up any more than we already have."

We kept pishing without success, and Cooper was growing frustrated. "The birds are right in here and we ought to be able to find them with some ease," he muttered, embarrassed that his vaunted scrub-jays failed to show. "On behalf of the 2A family, I apologize they have decided not to come out this morning. It's possible that they're just more interested in tending to the nest and eggs or young or whatever they may have, hopefully, and they're not willing to come out right now. So that's a good sign. So even though it's not good for us, it's good for the scrub-jays. This is normally a very gregarious group. They normally come out all five at a time. I don't even hear any calls right now."

We continued down the sand track to the next territory. We waited. "No scrub-jays here either," Cooper said, shaking his head.

On our way to the third territory, we passed a birder with a walking stick dressed in a jaunty khaki shirt and Bermuda shorts with binoculars hung on his neck. "Good morning!" he shouted before stepping to the side of the sandy track, waving us past.

"The scrub-jays are the main reason people like him come to the park," Clements said, waving back. He said that keeping scrub-jays around is not just a conservation issue; it's a political and financial one. The state money the park gets is based partly on the number of visitors. Declining scrub-jays means declining park attendance.

"It's amazing to see folks like this come into the park," he said. "Most birders have their life list they're working on. They're excited to find the Florida scrub-jay and add it. 'Oh, my God, I found the scrub-jay and the jay came right over to me!' I'll hear them say. People are just amazed at how friendly the birds are. A lot of people are thinking, 'Oh, I've got to trek out five miles and all this stuff.' But, no, you can be driving down the road in the park right off the main drive and have a chance to see the Florida scrub-jay up on the power line."

Clements stopped the truck at the third territory. We got out. Cooper turned suddenly and motioned us to be quiet. "Now, that's the scrub-jay," he said, pointing in the direction his ear was turned—a sharp *shreep*. The bird was perched in a tall slash pine twenty yards distant. A short while later, a sentinel landed in a tall pine snag on the opposite side of the road.

Cooper said that three years earlier, the female of the pair had died. "The male up in the snag there, he was sitting up in the tree after his mate died and he was just calling and calling and calling for the next couple days. Not long after, a lone female from a neighboring territory came over and joined him. They've built nests. This is a pair that knows how to breed, knows how to build a nest and raise young; they're just not doing it," Cooper said. "Why, is anyone's guess."

"Where's the female?" I asked.

"Well, it can be a good sign that we're not finding her. She might be sitting on a nest right now. We can only hope. The male on sentinel duty on top of that snag sort of implies that maybe the nest is over there."

Cooper was happy that we'd seen the scrub-jays. On the drive out of the park, we swung by 2A again in the hope of seeing that family. Still no one showed up.

Cooper confessed that he'd earlier spotted what he thought was a large hawk in a tree in the territory. "That might have been part of the reason the jays are hunkered down. I didn't say anything because I thought maybe it was a snag—part of a dead tree branch that can look a lot like a hawk, at least from a distance."

"Oh, yeah, the elusive 'snag bird,'" Clements added jokingly.

"We have a lot of those," Cooper replied. "Branch-like things that look like birds."

"The park wants to do everything it can to hold onto the few scrub-jays still here," Cooper said. "But it's just a matter of time before all the birds will be gone from the park."

Cooper said there was talk of translocating scrub-jays there from other areas, but the state wasn't having it. There was not enough habitat to support an ongoing population, and with little money to go around, it would be an unwise investment. As one Fish and Wildlife official summed it up, "Why would we put more scrub-jays there? It's a dead end."

7

DEATH BY
A MILLION NICKS

Like Sandy Cooper at Oscar Scherer State Park, Jon Thaxton has painfully witnessed the long decline of scrub-jays in Sarasota County. A former scrub-jay researcher, past county commissioner for twelve years, political provocateur, and unapologetic environmental warrior, Thaxton, sixty-one, has fought for more than thirty years to save the Florida scrub-jay in the county where his family has lived for five generations. Thaxton takes the loss of the scrub-jays personally.

Born near Osprey, not far from Oscar Scherer, Thaxton came from a real estate family and began working for the family company when he was fourteen. That year, he helped found the Ecology Club at Venice High School. He became intimately familiar with the relentless buying and selling of land, the escalating demand for and building of new houses and neighborhoods. At the same time, he understood the aspirations that brought people to what was once lovely countryside. Thaxton played a major political role in expanding Oscar Scherer in the 1990s and has worked to protect dozens of smaller parcels of scrub throughout the county, including the eponymous 287-acre Scherer Thaxton Preserve, located just east of Oscar Scherer State Park.

"We didn't save the scrub-jay in Sarasota County, and that's just a huge disappointment and sadness, but efforts to save it meant we saved many hundreds of acres that would otherwise have been lost," he said. "House by house, lot by lot, the loss over decades added up."

"Death by a thousand cuts?"

"More like by a million nicks," he said.

Today Thaxton is senior vice president for community investment at the Gulf Coast Community Foundation in Venice, Florida. In 2016, I met him at his office along the Tamiami Trail. We climbed into his car, and he said he wanted me to see some of the old scrub-jay haunts around the county. "Maybe we'll even see a few scrub-jays."

Thaxton said that in the early 1900s, Sarasota County had the largest concentration of scrub-jays on the Gulf Coast. Even by the 1920s, the birds had begun to decline in places while holding their own in other parts of the county. As late as 1954, they were commonly seen in some areas. But by the 1970s, only one large concentration remained, in South Venice. "Then houses gobbled that up," he said.

As we headed on the Tamiami Trail toward South Venice, where he thought we might be able to find a patch of scrub or two, Thaxton suggested we stop by a few remaining scrub islands elsewhere. He turned onto Airport Avenue East and then headed over to the coast to Service Club Park.

Thaxton said that in generations past, numerous patches of scrub throughout the county were close enough together that they functioned as a system. Oscar Scherer was one of the main islands in the archipelago of scrub that covered the county.

Then patches got smaller, fewer, and farther apart. "Connectivity was lost," he said. Without a connection to the larger landscape, a single, healthy island of scrub as in Oscar Scherer couldn't support the birds. Without connectivity, even well-maintained patches of scrub become a shattered mosaic. "The birds became isolated, the population dwindled, and the birds winked out," Thaxton said. "There was a time not too long ago when enough stepping-stones could have been preserved in Sarasota County. Now it's just a few fragments."

After a short drive, we pulled into the sandy parking lot of Service Club Park. Just south of that was Sharky's on the Pier, a well-known local hangout. Across the road, we could see Venice Municipal

Airport. Houses bordered the park to the north, and US Coast Guard Auxiliary Flotilla 86 hemmed it in on the south. We got out of the car, ducked through some brush, and entered the scrub.

"This is classic coastal scrub," Thaxton said. "It obviously hasn't had any fire pruning in a number of years," he continued, pointing out the tall brush. "This is the only place left on the planet where scrub-jay territory touches the Gulf of Mexico." He said the birds don't visit there very often now. None had been seen there between 2008 and 2013, and only two during 2016, according to the Audubon Jay Watch census. We didn't see any either.

Still, it was an enchanting few acres right on the beach, covered in a variety of shrubs presided over by native slash pines with conical tops reminiscent of bonsai trees. The sky was incredibly blue that morning, its azure depth intensified by sculpted white clouds on the horizon. The deep green of the slash pines, the brilliant white sand, and the turquoise horizon merged into a tropical painting anyone would love to hang behind their living room sofa. Picturesque pines grew down almost to the high-water mark, an intermingling of surf and scrub that has all but vanished along the Gulf Coast. Among the loveliest of Florida landscapes, coastal scrub has become the rarest.

We walked a narrow sandy trail into the few acres. "This needs to be burned, but it isn't at all," Thaxton said. "Unfortunately, what's going to happen one day is we're going to get a kid in here smoking a joint or a cigarette, and it's going to catch this pine straw on fire, and the whole thing's going to go up, and instead of pruning the pines it'll kill them, and that's too bad. This is really a very, very unusual habitat."

We watched two bald eagles flying in tandem just offshore. "Showtime. They're pair-bonding right now," Thaxton said. Then he pointed out a native winged sumac (*Rhus copallinum*) and said that aboriginals used the berry to make a dye to color their garments. Not far from it, he pointed to an invasive Brazilian pepper. The exotics usually don't take root in healthy scrub that has a full complement of native plants. But lack of fire gives the invasives a foothold.

Thaxton pished, pished, and pished for scrub-jays. "Don't hold your breath," he said, barely able to catch his. "A long time ago, there were nests throughout the area. But there hasn't been one in these parts in a very long time. That's why I wanted to bring you here. Before it's gone."

We returned to the car and headed south past Runway 31 at the airport. "There may be a family or two in the scrub remnants at the end of the runway," he said. "And over there, there's another patch of scrub at a retirement facility just across the Intracoastal Waterway. I don't know if there are birds still there. I don't think so."

A few minutes' drive south of that brought us to the Caspersen Beach scrub, fifty or sixty acres of fragmented habitat bordered by a golf course that was squeezed into the right angle formed by Runways 5 and 31. We walked a little farther inland. Much of the area was overgrown. All of three birds had been spotted in the area over the previous couple of years. "They won't be here for long," Thaxton said. "It was all scrub a few decades ago."

Another few minutes' drive south brought us to Shamrock Park, where six birds still braved the scrub patch quartered by access roads and gutted by four tennis courts and two basketball courts. We didn't get out of the car.

"These places," he said, referring to the visits we'd just made, are "the remnants of a necklace of the scrub that once ran along much of this part of the coastline.

"Scrub-jays have been lost everywhere," Thaxton continued, "but the big story is South Venice. It probably had the largest scrub-jay population on the southwestern coast of Florida. The scrub was on just these incredibly massive deposits of sand that stretched from the Gulf of Mexico all the way down to Charlotte County. There were so many scrub-jays, you didn't have to look for them. It was so noisy and plentiful that a blindfolded person with earplugs could go through there and find the scrub-jays.

"It wasn't just a continuous, huge patch of scrub. It was the most ideal scrub habitat in Florida for scrub-jays. It was that sweet spot that the Florida scrub-jays absolutely love, with the configurations of the oaks, and it was in the right place, where the fire regime happened in just the perfect frequencies for Florida scrub-jays. It was big, and so you had the chance there for a lot of Florida scrub-jay territories and saving the Sarasota County population."

Talking about it pained him.

"All this area near the coast was platted for development in the 1940s. Development kicked off in the 1960s, and the 1970s and 1980s wiped it out big-time. It was replaced with classic suburban sprawl, crappy, crappy, just really dumb. It took twenty some-odd years, twenty-five, maybe thirty years, because the development of the homes was very slow and the blinking out of the territories was very slow. For that entire period, people who moved to South Venice were thinking, 'This is great. We have Florida scrub-jays here and they're happy and healthy, and I can see them while raising my kids.'

"I'd rather somebody had come in and just shot the birds and thrown them into the street or had them soaked in oil so we could have taken pictures of them as individuals dying that way, as opposed to the species dying out very slowly like they did, with nobody noticing.

"The Florida scrub-jay has one of the most remarkable senses of place, and by place I mean almost like a global positioning system that says, 'This is the place, and this is where I'm supposed to be. This is where my father lived; this is where my grandfather was.' There's pretty decent evidence that many of these Florida scrub-jay territories have been in the same male lineage since before Europeans ever got here. I mean hundreds and hundreds and hundreds of years because of how they passed these territories down. When you come in and destroy the landscape, the birds stay, and they're forced into these behavioral changes and stay until they die."

After twenty minutes of driving, we still hadn't managed to find even a trace of scrub, and Thaxton was growing frustrated. "I've seen some around over the last few years," he said, ducking his head beneath the car's sun visor to look for any signs of it. "You just have to find it. Sometimes you can see little bits at the edges of people's yards until they plant it over with bushes or grass." He was sure he could find a patch of scrub somewhere, mentioning a place he had in mind. "Did I turn too early? I know it's around here somewhere."

We turned right and left, up Flamingo and down Palmetto, on to Heron and past Ponderosa. "We'll find some scrub in here somewhere," Thaxton repeated. "When I was here about five years ago, a few patches hadn't been converted to grass."

"Wait—I think there's some!" he exclaimed, hitting the brakes. "You can see it next to their yard. Can you see the little patch of dense oaks and the white sand? That's scrub."

He pondered it for a while, shook his head, and said, "Federal protection here was almost laughable at the time. The scrub-jay didn't stop development of any property. There was always a way around it. That's why the scrub-jay's going extinct."

By law, clearing occupied scrub required a permit from the US Fish and Wildlife Service. Thaxton said he couldn't think of a single case in Sarasota County in which Fish and Wildlife didn't eventually meet a request for a permit. He admitted that inaction by the county government didn't help. As a Sarasota County commissioner from 2000 to 2012, he was in the know.

To clear land occupied by scrub-jays, a builder or developer needed the euphemistically named "incidental take" permit, or an "Oops, I wiped out an entire scrub-jay family by building a house on top of it" waiver. But the application could be daunting because a biological survey and a Habitat Conservation Plan, or HCP, were required to accompany every application. Getting a permit also required mitigation—buying and managing comparable habitat elsewhere to make up for what's to be lost. The process could be prohibitively expensive for individuals, and in complicated cases the approval could take years. But there were easier, less expensive workarounds, including a mechanism whereby individuals could use a regional HCP and pay a small fee into a pool to meet the mitigation requirements.

Still, for at least a decade after the bird's listing as a federally threatened species in 1987, Sarasota and Charlotte Counties mostly ignored the legal requirements and allowed unbridled development of scrub-jay habitat, with or without permits. Fish and Wildlife did little to enforce the law until the early 1990s. By then, untold amounts of scrub-jay habitat had been destroyed.

Faced with newspaper accounts of blatant, illegal clearing of hundreds of acres of jay-rich habitat in the central part of the state in the early 1990s, Fish and Wildlife was forced to clamp down. Around 1992, it sent a letter to the Sarasota and Charlotte County governments warning that disrupting scrub-jay habitat was tantamount to illegal taking, or harming, of a protected species. Penalties could be severe, including hefty fines and, about as often as it rains in Death Valley, actual prison time.

"The 'smoking gun' letter was a complete shocker," Thaxton said, "because 'taking' usually meant killing birds directly by shooting or capturing them. But according to the new determination, the scrub-jay was so utterly dependent on specialized habitat that disrupting the scrub itself was tantamount to killing birds and was illegal unless the developer or landowner got an incidental take permit from the Service."

In response to the letter, Sarasota County officials posted on the county's website a map of scrub-jay habitat remaining around South Venice and elsewhere in the county. The map was based on Thaxton's own earlier surveys from when he worked as a researcher. It included more than a thousand acres of potential scrub-jay habitat. The website instructed county residents who planned to develop land where scrub-jays were present to call the US Fish and Wildlife Service, which was like telling aspiring tax cheats to ring up the Internal Revenue Service.

"This led to total hysteria," Thaxton said, "even though very few areas in the county still had scrub-jays. South Venice was one of them. It was a hot spot. It still had hundreds of undeveloped lots with scrub-jays. It was in the middle of development."

"During the hysteria, realtors moved in like they were first responders and warned people that it would cost thousands of dollars to prepare an incidental take permit and would take years to do the needed studies. It was total misinformation," he said.

According to an article in the *Sarasota Herald-Tribune* in 2005, land speculators warned owners to unload their "worthless" properties because scrub-jays were located on them. In some cases, landowners were offered as little as $1,200 per lot, far below the market value of $45,000.

Spencer Simon, a Fish and Wildlife supervisory biologist in Florida at the time, explained to me, "It was a big freaking deal, that letter. It's at the heart of it all. We got word that Sarasota and Charlotte Counties were issuing building or clearing permits to land we knew scrub-jays lived on. The Service informed them that this could be illegal under certain provisions of the Endangered Species Act.

"Elderly people who'd bought land, planning to retire, were being told by neighbors their land was worthless. I got letters from widows and widowers who said they were on a fixed income. People were telling them the federal government had devalued their land, which just wasn't true. I spoke to a meeting in Charlotte County, and one of the commissioners got up and threatened to take me outside and strangle me. I was appalled."

The *Herald-Tribune* quoted Simon at the time: "The hysteria is unfortunate. We haven't denied any permits in Sarasota County. As long as the applicant is willing to follow the federal guidelines, they should have no problem developing their land."

"After getting the letter, Charlotte County threw it in the trash and kept issuing county permits to anyone who wanted them," Thaxton said. "Sarasota County did it differently."

Technically, people with scrub-jays on their land needed to apply for an incidental take permit and create an HCP whether they planned to develop two thousand acres or a 125-by-75-foot lot. Although a company might have the money to hire a biologist to create an HCP, it was impractical for the owner of one lot. As a county commissioner at the time, Thaxton believed that with careful planning, development didn't have to mean extirpation of the scrub-jay from South Venice, let alone Sarasota County. He pushed for creating a county- or region-wide HCP and establishing community funds that individual builders could pay into as their form of mitigation. Each could put five hundred dollars or one thousand dollars into the mitigation fund. These funds could then be used to buy significant tracts of scrub elsewhere that would be permanently managed and protected. This would remove the onus from individual builders while centralizing planning for the scrub-jays' future. In other words, with the proper oversight, HCPs could work practically and politically.

Thaxton argued that "it was biologically meaningless to do an HCP on a single lot anyway." He was quoted in the *Sarasota Herald-Tribune*: "The county recognized that if our citizens are paying to prepare these individual plans—a cumbersome and almost useless process—wouldn't we all be better off if that money went into a fund that can have real benefit?"

"But the county eventually shot down the idea of the HCP before even looking at it," Thaxton recalled. "They didn't want any restrictions on development."

"I was the head of that group working on the HCP," recalled John Fitzpatrick, a prominent scrub-jay researcher and executive director of the Archbold Biological Station in Highlands County at the time. "And we worked very hard to give the county what it needed to move the HCP forward. They voted to reject the whole process." Both he and Thaxton were beside themselves.

In neighboring Charlotte County, the outcome was the same. In 2005, the county proposed such a plan for eleven thousand acres of identified scrub habitat. County commissioners voted that down too.

As we pulled away from the tiny patch of scrub he'd spotted next to the yard, Thaxton shook his head and asked bitterly, "So, what was the point of the Endangered Species Act if there was always some way to weasel out of it? South Venice was probably the last real chance to save a self-sustaining population of scrub-jays in Sarasota County and southwest Florida. Ultimately, the law didn't stop anybody from developing anything."

If South Venice was a gaping gash left in the Sarasota population of scrub-jays, most of the damage was inflicted, as Thaxton said, with a "nick here, there, and everywhere." The county had a long history of willful neglect.

In 1992, Sarasota County commissioners had voted to extend Pine Street, a main north–south thoroughfare through Englewood—right through scrub-jay habitat. The $9 million project would obliterate the homes of ten birds. Commissioners cited the same rationale for countless other road "improvements" in the county—improved traffic flow, reduced commute times, and a quicker hurricane escape route.

According to Thaxton, "the county had been aware of the birds there as early as 1988, when planning for the extension began. That was five years before the vote. There was time to easily redesign the extension and bypass birds. But commissioners said they'd already invested $1.5 million in planning and said any modification would be a 'waste of money.' As far as I'm concerned, that money was wasted anyway because they didn't use it to plan. They just used it to ram the project through."

Thaxton, who by then had become the conservation chairman for the Sarasota Audubon Society, continued, "The commission's failure to plan properly at that time was its own fault, and if redoing the planning would cost the county, that was the price of poor planning, not the price of saving the scrub-jays."

He said that if the county needed to save money, it could have easily forgone the extension of Pine Street. "An evacuation route already existed, and reducing the commute times was purely discretionary. What the commissioners really meant was, 'We'll spend nine million dollars on Pine Street, but we're not going to spend a dime on the bird.'"

The county's own development plan mandated protection of scrub-jay habitat. Faced with the fact that the Pine Street extension would violate the development plan, John Wesley White, the county administrator at the time, had a simple solution: legally waive it. White argued that the habitat was marginal and that there were only a handful of scrub-jays there anyway.

People on both sides of the proposed extension pleaded their case in the local press. "By expanding Pine Street, Sarasota County will destroy prime, populated habitat of the scrub jay, a rare and unique

Florida bird species," wrote Emilie McAlevy, a Longboat Key resident. "To whom is it worth the eradication of a small, stable population of special, admirable, interesting birds unable to cope with development pressures? Is blazing a minor road extension more important?"

David Atwater Jr., in the Our Land! column of the *Sarasota Herald-Tribune*, took a different view, suggesting that the birds' predicament was partly their own fault. "The scrub jay will not readily move to a new location . . . for this reason we are losing our population. . . . The intersection of Dearborn and Pine streets is no longer a viable home for our Florida scrub jays—the favorite bird for many of us in Southwest Florida. Here is a case of should we protect ten birds that are probably doomed and not likely to reproduce at the cost of hundreds of thousands of dollars."

As Thaxton and I approached Pine Street, he stopped the car. It was a busy road, and we didn't dare get out. But he leaned over to point out the scene of the crime.

"Right over there, that's where the nest was. I wrote the county a letter explaining that this is not the time you want to be screwing with this extension. The county commissioner said to hell with it. I had no idea they were going to be so foolish and send a bulldozer out there the next week."

When Thaxton went to visit the site a few days after sending the letter, "it was too late. The bulldozer almost seemed to have targeted the nest tree and gone right over it. It was nesting season, and I could have shown you the broken eggs. I called the county commission, and I copied in the transportation department. The county got in deep shit.

"The US Fish and Wildlife Service demanded that the county mitigate the situation by buying and managing scrub elsewhere. If they were going to mitigate, I was the person they had to come to because I knew all the scrub-jay habitat in the county. I'd mapped it three times. I got to pick a choice piece of land where there were still scrub-jays. It was a beautiful piece of scrub that is now a preserve in South Venice." It's called South Venice Lemon Bay Preserve.

Thaxton said the roadwork was just one among many similar activities at the time eating away at the scrub-jays. "When the Pine Street extension was started in 1992, about 350 scrub-jays remained in Sarasota County," he said. "That was a 90 percent reduction over the past century. By the time the extension never was completed, there were fewer than ten.

"Yes," he said. "You heard me right. The county commissioners canned the Pine Street extension."

Although the county's remnant scrub-jay population goes up and down marginally, the trend is unavoidable. As one senior Fish and Wildlife official told me, "It's so bad we don't even bother to include the Sarasota birds in our planning anymore."

According to Thaxton, "the saddest part of this story is that it was avoidable. The Florida scrub-jay in Sarasota County could have easily been preserved without stymieing the industries that were motivating its demise. We still could have had citrus and scrub-jays. It's not even citrus anymore because they planted it in an area where it was going to freeze. Duh! That was a real bright move. Even down here, many of the subdivisions and things could have been constructed in a way to preserve the most strategic scrub-jay areas, and we could have had both. There were people who were saying how to do it correctly. It's so frustrating. It was stupidity and greed."

We started to drive back to Thaxton's office, but he seemed hardly able to bring himself to leave without seeing a scrub-jay, so we swung by the South Venice Lemon Bay Preserve, a 222-acre parcel with a few areas of scrub and scrubby flatwoods. "There used to be quite a few birds here, but now no more than a dozen or two are left," he said. Ten birds were spotted there in 2016. Although we didn't see any, it seemed a comforting thought for him just to know they were still there.

Ever more determined to find one of the birds, Thaxton drove us down a road not far from Scrub Jay Court, an area of exclusive homes, and then toward the Manasota Scrub Preserve, 155 acres where several scrub-jay families lived among the surrounding suburbs. Thaxton pished as we walked along the sand road into the scrub. Almost immediately, two scrub-jays appeared, then two more in a pine snag to our right, one of which flew over us to the second snag across the road.

"I told you there were still scrub-jays in Sarasota County," Thaxton said as we got into the car and headed back to his office.

8

⌒ ꙮ ⌒

THE SCRUB-JAYS
OF BONE VALLEY

If Sarasota County is where scrub-jays went to die, neighboring Manatee County is where they lived to fight another day. They were lucky—or at least that was part of their success. Much of Manatee occupies the higher, scrubbier DeSoto Plain, while low-lying Sarasota is wetter and has less scrub to begin with. Much of Manatee's inland scrub lies beyond the initial clutch of sprawl, whereas many of the coastal scrubs and former haunts of the scrub-jays in Sarasota lay in the path of coastal development.

Even as citrus growers moved into Manatee's scrub-laden interior in the 1960s and 1970s, the scrub-jay lucked out. Elsewhere in Florida, growers cleared the higher scrub for oranges and grapefruit while leaving the lower surrounding flatwoods mostly intact. But in Manatee, the opposite often occurred: growers left a lot of scrub intact while clearing the flatwoods to get at the better soils. But if geography has generally been the happy destiny of scrub-jays in Manatee, geology has also been its nemesis.

Bone Valley, about thirty-five miles northeast of Sarasota, is a belt of buried phosphate clustered around where the corners of Manatee, Hillsborough, Polk, and Hardee Counties join. Phosphate, the source of phosphorus, a main ingredient of fertilizer, is in big demand. Plants need phosphorus to synthesize proteins for growth. The mineral is almost as important to plants as water. Required by living cells, phosphorus is a chemical building block of life. The green revolution would probably have failed without mined phosphate, and every one of us has eaten food grown with the mined mineral. Mined phosphate, for which there is no substitute, has become indispensable.

The phosphate deposits of Bone Valley were formed when seas covered the Four Corners area of Florida millions of years ago and the skeletons and waste products of sea animals and phosphorus from seawater settled on the ocean floor. Over the millennia, the veil of deposits grew into layers, the layers into strata, and the strata into a geologic formation. Largely an animal by-product, phosphate could be considered a gift from their world to ours. And it's a gift that just keeps on taking.

More than a century of mining across Bone Valley has destroyed vast acres of scrub, forest, and wetland. According to the Florida Department of Environmental Protection, "of the commodities mined in Florida, phosphate mining is the most land-intensive, disturbing between 5,000 to 6,000 acres annually." The largest mine covers nearly one hundred thousand acres, while the twenty-seven phosphate mines in the state have claimed nearly a half million acres. One of the world's largest phosphate mines covers fifty-eight thousand acres at Four Corners in Bone Valley, an area five times the size of Manhattan.

As the largest landowner in the Four Corners area, the Mosaic Company also came to be one of the largest "owners" of scrub-jays in Hillsborough and Manatee Counties. For that reason, there was a lot of concern in the mid-1990s when the company revealed plans to mine fifty thousand acres in the southeast corner of Hillsborough County and the northeast corner of Manatee. A sizable portion of it was scrub-jay habitat.

In the late 1990s, the Mosaic Company applied to the US Fish and Wildlife Service for an incidental take permit (ITP) related to digging up phosphate at the Four Corners mine. To get permitted for such a big operation, the company had to submit a Habitat Conservation Plan (HCP) along with the incidental take application. The

ITP addressed the best way to mitigate the loss of the large number of scrub-jays in the proposed mining area. This would be the biggest application ever submitted for Florida scrub-jays—and eventually approved.

To prepare the permit application, Mosaic turned to the environmental consulting firm Quest Ecology, in nearby Sun City Center, to survey, band, and monitor scrub-jays in the proposed mining area. Quest consultant David Gordon had conducted earlier surveys in the region and found a few small, isolated groups of scrub-jays. With or without the mining, Gordon had held out little hope for this Hillsborough-Manatee population. The birds were scattered, and their habitat was mostly overgrown. He feared they would eventually wink out as did the Sarasota County populations.

Reed Bowman, a scrub-jay scientist at the Archbold Biological Station in Venus, Florida, and a member of the Florida Scrub-Jay Recovery Team, had helped the team analyze the viability of different scrub-jay populations throughout the peninsula, including the population in Manatee County and neighboring Hillsborough County. The analysis concluded that, despite the large number of scrub-jays in the population, it had a high risk of extinction because the birds were isolated across many small patches. That was before Gordon discovered the twelve families on Mosaic land. With this bonanza, the question then became how to best preserve them in the face of the proposed mining. Whatever the strategy, Bowman was pessimistic.

"Even though this site had a relatively large number of birds, its extinction probability was very high because the populations were all small and isolated, and reversing this could be difficult," Bowman said. Mosaic had limited choices. It could leave the birds where they were and mine around them or try to move them. Based on computer modeling, Bowman predicted that if left where they were, the birds had only about a 10 percent chance of persisting. Even if Mosaic decided not to mine the area, overgrown habitat and inbreeding would eventually doom them. That left moving them as the only serious option.

Translocating all the birds would be a drastic step—like a kind of ecological organ transplant—and it had been tried only once before with the species. That attempt didn't work so well. Figuring out how to do it right could take years because Gordon would need time to move a few birds at a time and in different ways to see which worked best. They could move them by age-class, such as one- and two-year-olds, and see which age group fared best. Such experimentation would take time, and Mosaic wanted to mine right away. But, as Bowman saw it, there was little to lose by pushing Mosaic for time to move the birds.

"I suggested that Mosaic delay mining any of the areas with jays for ten years, during which time we would translocate the young scrub-jays and eventually move the entire family group," said Bowman. To just about everyone's surprise, Mosaic agreed.

The problem was that there was no good place to move the birds. Although plenty of scrub existed on Mosaic land, much of it had been marked for mining. The company didn't want to move the problem from one piece of its property to another.

But the company also owned a thousand acres in an area away from existing mining areas. It was too isolated to move the massive mining equipment to the site. This area had been leased to Manatee County as a wellfield that supplied some of the county's drinking water. Mosaic's part of the wellfield was under a conservation easement. About seven hundred acres of the wellfield property was scrub, although it was in abysmal condition—so poor that only a single jay group lived there at the time. But if Mosaic would agree to restore this scrub and finance translocation of the scrub-jays, among other things, a deal could be struck. As Gordon put it, "To make it happen, the US Fish and Wildlife Service told Mosaic all it had to do was write a check to restore that habitat and move the thirteen family groups. It was one of the biggest 'incidental takes' the US Fish and Wildlife Service had ever permitted." According to Gordon, "instead, they wrote a check and put a conservation easement on a large portion of the wellfield." The check would also cover the entire cost of the translocation.

Burned and restored, the wellfield could host at least two dozen families. While that would fall far short of the number needed for a self-sustaining population, the wellfield happened to abut the largest preserve in Manatee County—the twenty-one-thousand-acre Duette Preserve. Although the scrub there was far from ideal, it was home to six family groups of scrub-jays. Duette had four or five thousand acres of potentially good scrub. If restored, the combined habitat at the wellfield and Duette could hold upward of 115 families—the likes of which southwest Florida hadn't seen in at least a half century. In theory, that could rival the current population on Merritt Island. And it would be genetically distinct from the scrub-jay populations elsewhere in the state.

But simply moving the birds to the restored Mosaic wellfield scrub would not be a long-term solution. Success really depended on a growing population at the wellfield and Duette spilling into outlying areas of scrub. Together, all these groups of birds would give Manatee a large and stable population to complement the other strongholds around the state.

By 2000, the restoration team had "put fire back on the landscape at the wellfield," Gordon said. "We took down the forested structure mechanically first, mainly with chain saws. Cut all that down. We tried to get the natural fire regime going again. Before long the place was good for scrub-jays." Following restoration, two more family groups established themselves at the wellfield on their own. That made for a total of three families.

In 2003, Gordon and his team trapped and translocated the first seven helpers from family groups from the Four Corners area to the wellfield. The following year, they moved six more helpers, then four more in 2005. That was a total of seventeen birds. But after the third translocation, the birds had still not settled down in the unoccupied portions of the wellfield during the first year. Few had bonded. Gordon worried that the translocation technique might be to blame. So in 2007, he got permission from Fish and Wildlife to allow him to translocate entire family groups instead of just helpers, figuring that would solve the problem.

In addition to moving family groups rather than individuals, Gordon shifted the translocations from February, when the birds become highly territorial as breeding season begins, to November, when territoriality and fighting are at their lowest. As Bowman put it, this change in timing "allowed the released birds to establish their own territories without having to fight intensely with the locals."

Between 2008 and 2011, thirty-one more birds were moved to the wellfield using this new strategy. Eighty percent of those birds established territories. Between 2008 and 2011, the number of family groups at the wellfield grew from five to nine. Seven new groups were formed on Duette Preserve by translocated jays or offspring of jays translocated to the wellfield. What's more, seven of the existing groups at the Duette Preserve had at least one translocated bird or its offspring from the wellfield. Many of them had been hatched at the wellfield or had migrated from poorer habitat at Duette. All of the birds were finding a place in the new population, and the project seemed to be meeting the benchmarks of success. And over the next several years, the population kept climbing.

It was a dramatic turnaround for the species, whose number had dropped by 20 percent on protected land and by 70 percent on private land in the two decades before. As it turned out, all this success could not have come at a better time for Mosaic. Over the years, phosphate mining around Florida had earned an environmental record bordering on the macabre. For every ton of marketable phosphate produced, nearly five tons of a mildly radioactive by-product—known as phosphogypsum—remains. This soup is stored in aboveground reservoirs called gypsum stacks. These massive reservoirs sit atop the state's brittle limestone mantle. In 2016, at a Mosaic mine about thirty miles east of Tampa, the weight of one reservoir crushed through the limestone, and the resultant sinkhole swallowed 215 million gallons of contaminated water, sending it toward one of the more permeable regions of the Floridan aquifer. Mosaic's delay in publicly revealing the disaster compounded the public relations disaster. Fortunately, by investing so heavily in conservation of the scrub-jay, Mosaic had also been investing in public relations.

In the wake of the disaster, the amiable scrub-jay rapidly became a star of the company's public relations campaign. One television spot at the time stated, "Fifteen years ago, scientists didn't think local scrub-jays stood much of a chance. Florida Phosphate changed that. To save these local creatures from extinction, we preserved a thousand acres of habitat and worked with environmental experts to relocate scrub-jays. Right here, right where they belong. Today, our scrub-jays are thriving, and we plan to keep it that way."

Still, the company deserves credit. It could have taken a shorter and faster route to meet the requirements for mitigating the taking of scrub-jays in the decade before. It didn't. The result is a population of scrub-jays that, with careful management and continued luck, could become a larger, growing anchor population.

Bowman and Gordon are optimistic about the bird's future in Bone Valley, despite a dip in the population a year ago. "That's biology," Gordon said. "The trajectory continues to look good. While Mosaic did fund and take the extra step to translocate the jays from its properties, it was and continues to be great habitat, as well as scrub, on Duette Preserve that has kept Mosaic in permit compliance. Without Duette, the successes we have seen and continue to strive for would never have been possible. Now, with these birds connecting with others from the Mosaic wellfield and at Duette, we've seen a six- or eight-group increase in a year. It takes a long time to get the habitat back. But look at the juveniles this year; that's a good sign."

According to Bowman, "the translocation was successful, and Duette made it even more successful because it allowed the population to grow even larger. The important take-home message is that you need a lot of space to grow a population that is self-sustaining."

But with Manatee County's current scrub-jay population less than 10 percent of what it once was, risks remain. Some biologists worry that signs of inbreeding may show up in the future because the population was isolated for so long and because the translocated birds were all from the same genetic group.

Still, the "larger population will be less prone to genetic drift," Bowman said, referring to the loss of different genes in a small population as individuals die. This can cause the population to "drift" toward a genetic bottleneck. What's more, if the overall population of the Manatee birds consists of mostly small populations with little connectivity, "genetics will be a problem unless we translocate birds in the future from a different population."

According to Gordon, "As long as the political will and funding remain, as well as the skill set and desire of land managers at Manatee County to continue their fire management program, Duette should continue to grow more jays and family groups. This is also true for the surrounding public lands where smaller populations of jays exist and move between all the public lands and Duette Preserve."

For now, Manatee County enjoys an increase in scrub-jays even as neighboring Sarasota County watches the last of its scrub-jays disappear. Sarasota did too little, too late. Manatee may have done just enough, in the nick of time.

9

LOVE OF LIFE

"Love the animals, love the plants, love everything," the Russian novelist Fyodor Dostoyevsky wrote. Fifteen hundred years before that, Saint Basil the Great referred to other creatures as "our brothers, the animals." Nothing could have better reflected these age-old sentiments in the modern era than a group of birders who had gathered one spring morning in 2017 to express their love and concern for the small group of scrub-jays in a patch of scrub in Bone Valley not far from Duette. They came bearing gifts of binoculars, spotting scopes, and field notebooks.

The dearly devoted who gathered that morning were part of Florida Audubon's Jay Watch Program, the purpose of which is to gather information on scrub-jays at the Moody Branch Preserve to be used to track their dispersal, plan the best areas for prescribed burns, and chart the overall progress of the birds. Up early, without pay, and working long hours, they had come not only to seek census data but also to express, in their own individual ways, their bond with the bird and their loyalty and concern for its future.

The 960-acre Moody Branch Preserve lies about five miles northwest of the Duette Preserve. By the time Manatee County bought the land in 2004, the northern part of the scrub had been turned into cattle range and farmland. Fortunately, the scrub-jays were able to escape into the five hundred acres of livable scrub on the park's southern side, where their descendants still live today. As Todd Mecklenborg, who leads the US Fish and Wildlife Florida Scrub-Jay Recovery Team, and I pulled through the gates of the park that morning, several trucks, with volunteers milling around them, were parked in the grassy clearing. It looked like a tailgate party awaiting a sporting event.

Every year, hundreds of Audubon Jay Watch volunteers with this citizen science program visit nearly seventy-five sites— Moody Branch is one of them—in nineteen Florida counties to check in on the banded birds. In 2018, various teams of volunteers logged nearly three thousand hours in training and field surveys around the state. While collecting population data was the main goal, the activity seemed part ritual, part group therapy, and part communal catharsis—a meeting of like-minded souls striving for something much larger than the quantitative measures that would be entered into a spreadsheet the following week.

As ornithologists throughout most of the 1800s knew, tracking the movements of wild birds was all but impossible until 1890, when an enterprising researcher began to identify individuals by putting unique bands on their legs. Within just a couple of decades, banded birds had helped ornithologists identify several flyways across Europe. Today, bands are an essential tool for studying birds, including the Florida scrub-jay. Without bands, a fraction would be known about their territories, family structure, or distribution. Little would be known about the movements of other species that migrate.

So many banded birds now fly around that an international network of government organizations and private associates coordinates data from hundreds of thousands of bands returned by hunters, citizens, and researchers. Hundreds of millions of data points have been collected. There's also a federal bird-banding clearinghouse. The federal Bird Banding Laboratory (BBL) at the Patuxent Wildlife Research Center in Maryland has archived around seventy-five billion banding records and five million records of "encounters" since 1920, with a hundred thousand more added each year.

"It's very critical nationally for demographic, mortality, and harvest rates for hunting of waterfowl," BBL chief Bruce Peterjohn said. While many of the banded birds wear only a silver federal band, lots of others wear various color combinations of leg bands that can distinguish individuals in groups. "These help estimate survivorship, local movements, and other information used to manage habitat for endangered species," he said. Unlike the silver metal bands alone, brightly colored ones can be seen by citizen-scientists with telescopes or binoculars who independently gather data and report it to the BBL. Volunteer data-gatherers are now an important constituency for the conservation of birds.

The birds at Moody were banded by David Gordon, who was working under a federal banding permit through his employer, Quest Ecology. Each bird gets one of the silver federal bands with a unique nine-digit number. Additional colored bands, which can be bought from private companies, are put on each leg. The scrub-jays often get three colored bands, which are usually read in sequence from the upper band on the bird's left leg to the lower band on the right leg.

Bands are like miniature license plates for scrub-jays. Lacking the distinctive individual markings of many other species such as the unique black-and-white patterns on the underside of a humpback's tail fluke or the white blotches on a killer whale, individual jays can be impossible to tell apart. In watching a number of unbanded scrub-jays darting in and out of the bushes, you could hardly tell if there were two or twenty, let alone who was who.

Soon after we arrived, the group of volunteers split into two parties. Mecklenborg and I joined the team that would cover the thirteen checkpoints in the south of the park; the others would canvass the partially restored north acreage. Our team included Audubon volunteers Mary Keith, an extension faculty member at the University of Florida's Foods, Nutrition, and Health Center in Seffner; biologist and avid birder Katherine Prophet; and Brendan O'Connor and Joshua Agee of Florida's Fish and Wildlife Conservation Commission. We got right to work and drove along the sandy road toward the first census point. Mecklenborg got out, climbed into the bed of the truck, and held up a recorder with a directional speaker that blared a series of scrub-jay calls. Then we waited for the wild scrub-jays to holler "Present." None did.

At checkpoint two, we saw only two distant scrub-jays, probably spooked by a red-shouldered hawk sitting on a pine snag on the horizon. For an hour, we went from checkpoint to checkpoint without getting close enough to a bird to identify the color of the leg band. At another checkpoint, Mecklenborg played the call.

"There goes one!" he said as a bird shot from the wiry branches of a twenty-foot-high oak on our left to a lower branch thirty yards to our right. It was joined by another while a third bird perched atop taller oaks. Wary and curious, they rocketed in flight and then peered cautiously from their perches.

"I got a female to the right," Mary Keith said, peering through her scope on a tripod. She said she knew it was a female because of the unique female hiccup call: *eic-eic-eic-eic-eic-eic-eic.* Only the females do that," she said. "I got a band here."

"Ah. Ah, Todd. Todd! Over your left, over your right shoulder!" I heard Keith shout as another bird showed up about forty yards to the right of the female, who was still perched atop the slender oak in front of us, about twenty yards inside the scrub. "Yes, another banded bird," she said, reading out the colors. "Silver over azure. Right leg I can't see. It looks like it's got a yellow on it, maybe," she said, peering through the scope.

A scrub-jay perched and preened on a slender branch atop the skinny oak while earthbound volunteers stood with binoculars and clipboards.

"Is it red over black and white?" O'Connor was asking for confirmation.

"No, red over black is on the right leg."

"Red white?" O'Connor asked, struggling to keep hold of the clipboard with the record sheet flapping in the breeze as he tried to get a shot of the bird with his camera, sporting a long Nikkor telephoto lens.

"Right leg," Keith corrected with an edge to her voice. "Come look."

"Red over black on the right leg," she repeated. "Silver over azure on the left."

And so the morning went as if grown-ups were playing "I spy with my little eye."

Mecklenborg's large binoculars, O'Connor's telephoto lens, and Keith's Bausch & Lomb were all trained on the birds in the large oak as they tried to get a glimpse of their bands. "It's a baby or juvenile in the tree," Keith repeated. "White over green is soothing a juvenile in the tree. It's got a flesh over something on the right leg. I can't see the other color."

"How do you know it's a juvenile?" I asked. "It was begging," she replied, not taking her eyes off the target. "It put its wings out and made them shudder. Or shook them. That's how I knew it was a juvenile," she said. Juveniles also have brown heads because they haven't gone through their first molt.

"There are five right here," Mecklenborg said, pointing to the thicker, shorter oak on the right. "There's one, two, three, four," he said, pointing them out with his finger. " . . . Five."

"No, there were six!" Keith corrected.

"We got one that was silver over bright green," Keith said.

"I thought it was white over green," O'Connor said with an Irish lilt.

"Now we've got a bright-red over silver on the right leg," Keith said. "Take a look at this. I don't see a band on the left leg," she said to O'Connor, handing her spotting scope over. "Look, Brendan."

"I think that's the one we're looking at," Keith said. "Is it green on the left?"

A few lonely longleaf pines broke the horizon, and the branches of one close by were hung with Spanish moss as long and lush as an Andalusian's mane. Sunlight glittered on the white sand.

O'Connor, still struggling with the clipboard, said, "Silver green on the left; does that make sense? We have that one," he said, meaning it was now marked on the record sheet.

"Silver green or white green?" Keith asked.

O'Connor looked up, waiting for an answer that never came.

Moving back to an earlier bird, he said, "I think the right has a dark band that makes it blend in with the leg."

"That one, in the right tree. Is that pink over . . . I can't see," O'Connor said.

"I think we've got this bird," he said. "Hot-pink over silver, left leg; azure over white, right leg."

The birds soon resumed their daily business, flying away or ducking back into the scrub. The Jay Watch crew were elated by the spectacle we'd just witnessed. We climbed back into the truck, and the team retold stories of the morning as storm trackers might describe their brush with a tornado.

As we stopped for the final checkpoint, I asked Mary Keith why she was so devoted to the Jay Watch census.

"For the data," she said, before pausing. "And for the wonder." She pointed skyward at the silhouette of a swallow-tailed kite gliding high above. After breeding in Florida, it would be returning to Venezuela for winter. Although these kites usually feed on snakes, lizards, frogs, and large insects, they've been known to swoop down and snag baby scrub-jays from their nests, Mecklenborg said. "They eat on the fly. Crazy!"

Keith continued to follow the path of the kite, this one far too high to be concerned with scrub-jays. More likely it was eyeing the hundreds of dragonflies, bronzed and gleaming in the sun, that were circling in a glittering vortex of metallic confetti a hundred feet above us.

"I was once driving through Micanopy, south of Gainesville, when I saw a group of kites dive-bombing a cloud of thousands of dragonflies," Keith said. She described how the kites zeroed in on dragonflies with such precision that they severed them right at the thorax, causing thousands of the shiny wings to flutter to the earth in a kaleidoscopic display of light. "Wow!" she said, pointing overhead again. "You just never know when something wondrous is going to happen." The kite overhead seemed to circle in ever-increasing synchrony with the forming eddy of flickering insects above us, as if timing a descent into the glittering swarm.

Soon the two teams reunited at the trucks and vehicles near the entrance gate of the preserve. They sat on tailgates or leaned against hoods. They shared "I remember one time when I was out looking for scrub-jays . . ." stories, scientific insights, and controversies surrounding the bird. Although many of the topics focused on habitat, Mecklenborg ventured that "Saving the scrub-jay would require much more than saving and managing land." With that, the conversation turned to climate change, the cars we drive, the homes we live in, and even the orange juice we drink, which after all, comes from groves that stand on scrublands once occupied by the scrub-jay. We are all implicated, one way or another, in tearing the fabric of the natural world. Perhaps the difference is that some people try, in myriad ways, to mend it.

The sentiment that animals are our brothers and sisters worthy of love, care, and affection has been expressed in our own age as "biophilia," a term popularized in the 1970s by psychologist Erich Fromm, who defined it as "the passionate love of life and all that is alive." A quarter century later, biologist E. O. Wilson repurposed the term to describe what he considered to be people's innate affinity for other species. Yet, all these notions of humans feeling connections with other species seem hopelessly lost in the world of vanishing scrub-jays and mounting extinctions. To join a group like the Jay Watch volunteers that morning—and to recall the dozens of other scrub-jay researchers like Mecklenborg—reminded me that there is still reason to believe.

PART THREE

THE LAKE WALES RIDGE

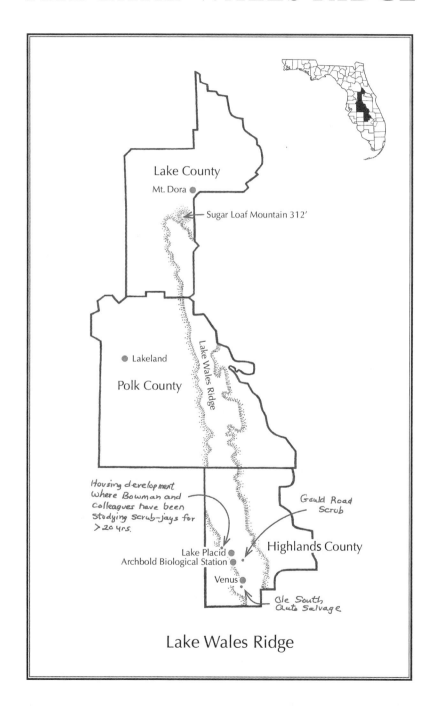

Lake County

Mt. Dora ●

— Sugar Loaf Mountain 312'

● Lakeland

Lake Wales Ridge

Polk County

Housing development
where Bowman and
colleagues have been
studying scrub-jays for
> 20 yrs.

Gould Road
Scrub

Highlands County

Lake Placid ●
Archbold Biological Station

Venus ●

Ole South
Auto Salvage

Lake Wales Ridge

IO

JOURNEY TO VENUS

Venus, Florida, an earthly outpost of about a thousand people in southeastern Highlands County, is a hodgepodge of old and new, rich and poor, past and future, sacred and profane. The town is home to the Venus Project, an organization that promotes futurist Jacque Fresco's vision of sustainable cities; a Methodist and a Baptist church; several major citrus growers, whose holdings consume an outsize amount of land in the area; assorted auto-repair shops; and one of the biggest car junkyards in the state, Ole South Auto Salvage. The only to-do item listed on TripAdvisor's Venus web page is the Archbold Biological Station, the Grand Central Station of research on the Florida scrub. Archbold's small cadre of scientists have published hundreds of scientific papers on scrub plants and insects and more than two hundred papers on the scrub-jay alone. Archbold is also the site of the longest nonstop study of wild birds in the United States. The study includes the meticulous annual mapping of every scrub-jay territory at Archbold since 1969.

Numerous world-class researchers have been drawn to Archbold by its location at the southern end of the Lake Wales Ridge, which holds one of the world's most interesting collections of endemic plants and animals. Formed by a series of elevated landforms left behind by high sea levels during different geologic ages, the Lake Wales

Ridge's seventy-five miles of white-sand uplands are visible from space.

The Lake Wales Ridge is a distinct landform within the larger central ridge, which begins just north of Ocala. These central highlands also include the Orlando Ridge, near the city of that name, and continue southward toward the more ancient landform of the Lake Wales Ridge, often referred to simply as the Ridge. Centuries ago, Florida scrub-jays lived throughout the sandy uplands along the entire length of these central uplands. Today they are concentrated on the Ridge, with the largest number of families residing in or around Archbold.

The scale of the scrub loss along the central ridge, much of it from conversion of scrub to citrus groves, is difficult to comprehend. In the 1960s, *New Yorker* writer John McPhee wrote in his book *Oranges* that citrus trees cover the ridge "like a long streamer, sometimes as little as a mile and never more than twenty-five miles wide, running south, from Leesburg to Sebring, for roughly a hundred miles. It is the most intense concentration of citrus in the world." Before the arrival of citrus and other development, it was the most intense concentration of scrub-jays in the world.

Historically, citrus growing targeted the dry uplands, so scrub bore the brunt of the clearing. Hoodwinked by all the romantic lore, many people came to believe that the orange was as native to Florida soil as the cabbage palm. But the main role of sand is to act as a hydroponic substrate to hold the trees upright while growers input the vast amount of fertilizer, water, and pesticides the Asian fruit needs to survive in the alien environment. The myth of Florida's "native" fruit of Eden has become so pervasive that the environmental costs of converting so much scrub to groves has largely been overlooked. The aroma of its blossoms so intoxicated the trees' admirers that state officials proclaimed the orange blossom the state flower in 1909, even as the orange tree was wiping out the scrub-jay and other true natives of the Florida soil and sky. Groves blanket so much scrubland of the Lake Wales Ridge that their geometric footprint is visible from space, and by 1990, more than two-thirds of scrub along the Central Ridge had been lost to citrus. The surviving scrub-jays there have splintered into small groups in the remaining fragments of scrub and are at high risk of dying out. Despite the bad news, the Lake Wales Ridge, according to Reed Bowman, a scrub-jay researcher at Archbold, has "the most connected network of scrub preserves in the state, and survival prospects for jays are better than most."

There are more reasons than jays for preserving the remaining scrub there. The Ridge comprises some of the oldest habitats in Florida. Like most other relict sand ridges in the state, it was left behind when high seas retreated. While high sea levels returned several times over the ensuing couple of hundred thousand years to reinundate most of the central ridge, the higher regions of the Lake Wales Ridge were not touched by seawater—and haven't been since around the time the scrub-jays arrived in Florida two million years ago. Given the Ridge's age and isolation, many species there had thousands of millennia to evolve undisturbed. Two federally listed endangered shrubs, the pygmy fringe tree and the scrub plum, are mostly limited to scrub patches on the Ridge in Highlands and Polk Counties. An even rarer shrub, Garrett's ziziphus (*Ziziphus celata*), with spiny zigzag branches, is one of the rarest plants in North America, and six known wild populations exist only on the Lake Wales Ridge. All told, nineteen federally listed species and numerous state-listed ones share this rare habitat with the Florida scrub-jay. Other still unknown plant and insect species likely await discovery.

With the sandy uplands having been under assault for more than a half century, nearly every acre of protected scrub that exists there today has been hard-won. Much of the credit goes to Archbold and the state, federal, and private conservation organizations the station helped knit into a broad-based network. Today the research station property is the linchpin of nearly fifty thousand acres of state-owned land and private conservation easements connected by a network of corridors and stepping-stones.

In the spring of 2016, I joined Bowman and scrub-jay expert John Fitzpatrick at Archbold, where they were conducting the annual mapping of the scrub-jays. I arrived at the station in the late morning, just as they were returning from the scrub. "This is my fortieth consecutive year mapping scrub-jay territories at the station," Fitzpatrick said gleefully. "I came down here as a college intern in the summer of 1972. In 1973 I started mapping the territories, and I haven't missed a year!"

After making our introductions in the lab—it was the first time I'd personally met Bowman or Fitzpatrick—we walked outside and got into the station's truck. Bowman drove, and Fitzpatrick rode shotgun as we headed out into the scrub. Spring was just arriving. Subtle and brief in this part of Florida, the fleeting season appears as a whisper and is quickly consumed by summer.

"Spring is one of the nicest times of the year in the scrub because the oak leaves turn color and fall off and new leaves sprout," Bowman said as he leaned over the steering wheel for a better view through the windshield. "There's more color out here right now." The normal khaki green of the scrub had taken on a bronze to reddish sheen as last year's leaves faded before being pushed off the branches by shiny new shoots.

"We're just a little short of the full leaf drop," Fitzpatrick said. Captivated by birds for most of his life, he often references seasons not by months of the Gregorian calendar but by the choruses of birds. "Another three weeks from now, that's when the bobwhites start to call," he said. Like Bowman, Fitzpatrick is enchanted by the scrub any time of the year—when it's mild, when lightning storms sweep through, and on those beastly hot summer afternoons when the sun's cauldron brings photosynthesis to a stop and nothing seems to move. But lots was moving now—the leaves in the light breeze, the clouds, and, imperceptibly, the swelling buds on the small oaks and other plants.

Many people see the scrub as a place only of blistering heat and prickly bushes. But the softness of the colors and the sheer grace of the land we rode through on this morning was enough to stir that almost primal sense of wonder usually reserved for grander habitats like rain forests and mountain ranges. As we drove, the mounds of bright rosemary bushes and pointy-leaved palmettos grew smaller as they receded into the white-sand horizon behind us. Ahead, parallel stands of pines stood stenciled against the blue sky. We passed ridges and swales, dry shrublands, oak woodlands, bayhead forests, and groves of hickory, pine flatwoods, sinkhole ponds, and ephemeral marshlands, all joining in a mosaic of shifting, flickering shadow and light.

I asked Bowman what he most enjoyed about being in the scrub.

"What's *not* to love?" he asked.

"So why *does* scrub get such a bad rap?"

"The inability to look past one's own nose," Fitzpatrick replied. "Unless you're in love with wildlands, it takes some level of experience with this habitat to deeply appreciate it. You have to understand this habitat from the point of view of a gopher tortoise to really understand its magic. For the public, it doesn't have the lush, beautiful high diversity and look and feel. It's hot, and it's sandy. It's got a bunch of characteristics that aren't charming. But this bird, anytime anyone gets to know it, they fall in love with the scrub-jay."

Before long, Bowman stopped the truck and we got out. A scrub-jay darted from the scrub and landed on the top of my head.

"They've been expecting us," Bowman said.

I was curious. "Why would any bird do that—just land on someone's head?"

"Besides for peanuts?" Fitzpatrick asked. "Maybe they evolved the trait to get a good vantage point, which they like to do, from the backs of camels, bison, or other large mammals that roamed this area during the Pleistocene." It's the same reason you often find sentinels perched in pine snags. A human head apparently fulfills the instinctive need and sometimes comes with a bonus of peanuts.

We climbed back into the truck and continued the drive down the sand track. Bowman pointed ahead to an embankment of distinctly yellow sand, different from the bleached white sand that scrub normally grows on. We pulled over again and got out, and no sooner had we placed our feet in the unusually tawny sand than Fitzpatrick leaned over and pointed out, at his feet, the yellow-sand species *Dicerandra frutescens,* or Lake Placid scrub mint. It belonged to the same family as the Titusville balm that Paul Schmalzer and I had seen a few months earlier at the Dicerandra Scrub Sanctuary in Brevard

County. Lake Placid scrub mint blooms in late summer or in the fall, but we were there to see new rosemary-shaped leaves just pushing out from the woody stems. Fitzpatrick pinched off a leaf and handed it to me. As I squeezed it between my thumb and index finger, the plant gave off a menthol scent, sweet but less fruity than that of the Titusville balm. Some have described it as a peppermint smell. Discovered in 1962, the endangered plant naturally covers no more than a few hundred acres in Highlands County.

Bowman said the scent released from tiny glands of the injured leaves repels insects. The aroma is strong enough to carry downwind for several yards. Among the dozen or so insect-repelling mint oils found in the plant is one previously unknown to science. Some species are naturally immune to the "repellant." One type of caterpillar eats the plant and, in turn, uses the juice for self-defense. When disturbed, the caterpillar vomits it up.

Bowman said that different species of oak and other plants, in addition to the scrub mints, tend to grow in yellow rather than white sand. Fitzpatrick reached for the bright, shiny leaves of an oak. "*Quercus myrtifolia*—a myrtle oak. It's a yellow-sand species and much denser, tougher, and grows more rapidly after fire. It's a really dense shrub," he said, reaching between the branches to make his point as the heavy foliage swallowed his hand.

It wasn't long before several scrub-jays appeared in a nearby bush. After so many years studying the birds at Archbold, Fitzpatrick and Bowman recognized them by sight.

"Is that the tame guy?" Fitzpatrick asked, referring to the one perching at the end of a branch near us. "He's normally tame but less so today," Bowman said. "That's his dad, above him there." Soon a third bird appeared, probably the young male's mother.

Scrub-jays are among the very few bird species that bond as families or engage in "cooperative breeding." In most other bird species, the young disperse as soon as they can fend for themselves, perhaps never to see their parents or siblings again. But the Florida scrub-jay is different. The male and female we were looking at would likely remain together for years, or until one died. Throughout this time, they would be accompanied by at least some helper offspring from the year before.

Parents may have up to a half dozen helpers, or "nonbreeding auxiliaries." While young females often go off to start a new family a year to two after leaving the nest, males may stay around, like the gregarious one we were looking at. These helpers spend most of their on-duty time patrolling the family territory and fighting off trespassers, protecting nestlings, and gathering food for them. What might appear to be an act of altruism or self-sacrifice is, in fact, pure self-interest.

A young male who heads out on his own, with little experience or territorial knowledge, is vulnerable to hawks and other predators. By remaining at home under an apprenticeship, a young bird also gains valuable experience in how to defend his parents' territory before having to defend his own. And there's the added benefit of possibly inheriting the territory from his aging parents. Furthermore, cooperative breeding may be a good strategy for living in patchy territories where food can be scarce and aerial predators ubiquitous. In such an unforgiving environment, it takes a family to raise a fledgling.

Fitzpatrick suddenly looked up into some taller trees by the road. "I was wondering if there was a pair of bluebirds around!" he exclaimed. We watched as a pair swirled about in arcing flight. He lifted his binoculars to view the top of a tall dead pine to see if there was a hole they might be nesting in. Then he quickly pivoted again, his binoculars still held tightly against his eyes.

"Look at those!" he said. "Pine warblers." He examined a flock of them and then let his binoculars slide from his hands and hang from the strap around his neck before shaking his head in wonderment. "They probably migrated from Georgia," he said as his index finger retraced their path through the sky. "So how about that? A pair of bluebirds and about twenty-five pine warblers!"

After the birds had passed, Fitzpatrick swept his arm toward the open horizon and said, "This seemingly vast expanse of land, it's just a remaining crumb left from times past. The citrus industry in the early part of the twentieth century and then the invention of the air conditioner in the middle part of the twentieth century, which vastly changed human migration into Florida, inexorably busted up the original distribution of scrub on the Central Ridge into a thousand little islands of various sizes. And only a few of those islands are large. This is one of them.

"When I first started coming down here in the 1970s, this southern section of the Lake Wales Ridge had a thousand scrub-jay families. Now there're only about four hundred. Archbold has only 120 or 130. And this is a relatively well-protected part of its range. So even in this, the population across the Lake Wales Ridge has crashed. Make no mistake. The scrub-jay is headed for extinction."

Bowman wasn't as dire. "Because it's headed for extinction now doesn't mean it will become extinct. But it will, unless more is done soon," he said.

Whatever their differing views on the future, both agree the scrub-jays are faring poorly, even on the Lake Wales Ridge. According to Bowman, between the early 1990s and 2010, scrub-jays on the Ridge suffered a 25 percent decline. Even in the best-protected areas, scrub-jays decreased from 554 families in 1985 to about 263 families in 2016. In 2011, about half of the scrub-jay populations on the Lake Wales Ridge remained stable or increased, and with increased burning and management, the other population could stabilize or even increase.

But the "stable" population doesn't mean the birds are not in trouble, according to Fitzpatrick. Currently, the birds at Archbold seem to be holding their own. Historically, birds from the smaller outside populations frequently migrated to Archbold, helping to stabilize the population there. But as these surrounding populations have declined or disappeared, so have the in-migrants. Without them, the Archbold birds are turning more to one another for mates. The gene pool is decreasing, and new offspring are becoming more inbred. While this hasn't reduced breeding success—yet—if it keeps up for, say, fifty years or so, it may well lead to genetic depression, or bottlenecking. Heavily inbred birds become less "fit." In turn, fewer eggs will hatch, and those that do will produce chicks that weigh less than healthy ones and die at a higher rate. The Archbold population would gradually decline. No amount of good local management of the scrub can make up for the loss of genetic diversity.

Conventional wisdom suggests that a large scrub-jay population on well-managed land will thrive. But as Fitzpatrick lamented, "We're beginning to see, even right here on this beautifully managed, beautifully protected fairly large island of scrub-jay habitat, the genetic structure is deteriorating because we're getting fewer and fewer immigrant breeders coming from adjacent habitats. They're heading toward a genetic bottleneck." It's no longer enough to just manage habitat. You have to manage DNA.

The findings at Archbold could have grave implications for scrub-jays elsewhere, according to Mecklenborg, leader of the US Fish and Wildlife Service's Florida Scrub-Jay Recovery Team, who had accompanied me to Moody Branch. "The problem is that the genetics haven't been studied anywhere but at Archbold. If it's happening there, it's probably happening just about everywhere else," he said, alluding to the slowly growing populations in Manatee County and the southern Atlantic coast population at Jonathan Dickinson State Park in Martin County.

This gotta-have-migrants epiphany severely undercuts the honored tradition of focusing on saving large populations of scrub-jays while triaging out the smaller ones. But now it's too late to save many of these suddenly important small feeder-populations. The bureaucracies of state and federal endangered species policy are unwieldy ships, and they aren't quickly maneuvered when trouble appears off the bow. Short of importing birds into the population, Fitzpatrick said, "the most we can do is restore habitat and close the gaps sufficiently so that birds will keep moving from one reserve to the next. And if you look at a map of reserves along the Ridge, we've done a decent job of preserving islands that are potentially close to each other."

According to Bowman, the plan of protecting enough land so the scrub-jay could naturally flourish and migrate seems a bit like a fantasy, in retrospect. "There will always be some kind of intervention," he said. "We know that's true because we will always have to manage fire because habitat patches aren't big enough to burn naturally. Similar genetic interventions might be needed. There's really not much we can do to rectify the genetic problem at Archbold unless that might mean actually physically moving birds."

Despite the success of translocation in Manatee County, moving birds isn't a panacea. Different scrub-jays aren't necessarily interchangeable from one group to the next. In theory, moving too many birds from one place to another could cause the opposite of inbreeding depression—outbreeding depression. Moving birds that are finely adapted to one area into a population adapted to another locale could potentially make the recipient population less fit. While some biologists are skeptical that the genetic differences among scrub-jay groups are big enough to cause this, the only way to find out is to mix them and hope for the best.

One begins to wonder at what point the wild scrub-jay becomes a captive species in an open-air zoo. Burning their habitat with drip torches or incendiary capsules dropped from helicopters is already routine. Scrub-jays are now being transported from one place to another in trucks because they can't fly far enough to link scattered scrublands on their own. This isn't the future most people are hoping for, but as Bowman suggested, this future has already arrived.

I I

LAKE PLACID

Several miles north of the Archbold Biological Station lies picturesque Lake Placid, one of numerous lakes and ponds across the Lake
Wales Ridge. Until the late 1950s, Lake Placid's shoreline was largely
undisturbed scrub, part of a band of more than six thousand acres of
jay habitat stretching for almost ten miles from Archbold north along
Lake Placid and neighboring Lake June-in-Winter. Old-timers can
still remember seeing scrub-jays throughout the area. Roads were few
and developments scattered. Aside from pastures and groves, the barriers to the birds' movements were largely natural, including wetlands
and lakes, high pine or flatwoods, hardwood hammocks, and open
prairies they refused to cross.

Drawn by mild winters, inexpensive land, and storybook ponds,
in the late 1950s and early 1960s retirees and others seeking second
homes away from Orlando, Jacksonville, and other big cities discovered this picturesque land of lakes on the Lake Wales Ridge, along
with the area's plentiful scrub-jays, an ever-present symbol of the
countrified lifestyle people longed for. By the 1970s, the flow of people drawn to the Lake Wales Ridge had become a flood.

The region's scrub began to quickly disappear. By the early 2000s,
agriculture, ranching, and residential and business development had
destroyed almost 75 percent of the Lake Wales Ridge's xeric uplands.
Of the three million acres of xeric uplands found before Europeans

settled on the Lake Wales Ridge, less than 450,000 survived. The vast majority of that remained unprotected, and despite the successful setting aside of 45,000 acres between 1985 and 2005—much of it not scrub-jay habitat—the rate of loss continued. The pittance remaining was home to some thirty species the US Fish and Wildlife Service considered endangered or threatened, many existing nowhere else. This included the scrub-jay and the Florida mouse (*Podomys floridanus*)—the state's only endemic mammal—and other species.

In 1985, Steve Christman, a conservationist and scientist working on contract with the Florida Game and Fresh Water Fish Commission, undertook the epic task of identifying all the significant patches that remained. Having done this—he found about twenty-five thousand acres in all—he was then tasked with prioritizing each puzzle piece to create a functional network that encompassed all the Ridge's endemic and endangered plants.

It would have been difficult to find a better person for the job. Christman had experience, knowledge, fortitude, and, perhaps above all, an almost visceral connection with the scrub. More than twenty years after he had produced his landmark report on the Lake Wales Ridge, he was still going strong, and I was lucky enough to accompany him on one of his forays into the world he knew as well as anyone alive. Friendly and conversational, as we walked through the scrub deep in Ocala National Forest located at the northern tip of the central ridge, he gave a running commentary not just on the plants but on every creature we came across. He'd even brought along a small rake to occasionally comb the surface in the hope of rousing an unusual lizard known as the Florida scrub skink that lives under the sand.

Back in the day, the *Orlando Sentinel* published a lengthy profile on Christman titled "Expert Documents Scrub's Demise as Bulldozers Rush In." "What I do for a living is document the extinction of our native flora and fauna," he is quoted as saying. "I can't stop it. I just document it."

The 1988 article continued: "Christman wears cowboy boots to keep cactus needles from piercing his feet. He buckles a compass around his wrist to avoid getting lost. . . . In his 6-year-old Volkswagen Rabbit, with 134,000 miles on the odometer, Christman packs a shovel to dig his car out of the scrub's deep sand, a rubber frog for company, and reference books." The profile also described Christman's discovery of a new plant species in a dump near Sebring. He called the new mint "one of the loveliest of God's creatures. If [I] hadn't found it, it probably would have just gone extinct, and no one would have ever known. . . . Life's a bitch, and then you go extinct." He planned to name it *Dicerandra salvata,* or salvaged *Diceranda,* but it ultimately got the name of its discoverer: *Dicerandra christmanii.* Throughout his three-year foray into the scrub, he traveled the razor's edge of extinction. "On the very day . . . Christman discovered the endangered scrub plum growing on a vacant lot in Orange County, the county commission rezoned the land for a subdivision," the article stated.

In 1989, Christman presented his findings to the Florida Game and Fresh Water Fish Commission and numerous others who had gathered at Archbold for an emergency conference on the Lake Wales Ridge scrub. "Development on the Lake Wales Ridge is proceeding unabated and dozens of significant scrub sites that I personally have surveyed in the past three years have already been destroyed," Christman stated in his report. "Others . . . are being developed for housing or agriculture as I write these words. Others will be root-raked to destroy the scrub vegetation and thus avoid the potential for government intervention and land use restrictions."

Christman described how one land company, Lykes Brothers, root-raked "one of the most diverse and most well-known of the ancient scrubs" at Josephine Creek, South East US 27, near Lake June-in-Winter, and that "it has lain fallow ever since, and the scrub plants have not returned. . . . The well-known ancient scrub on the south side of Josephine Creek where it passes under US-27 is home to no fewer than 11 [rare] scrub plants. The part east of the highway was root-raked in 1986 and now supports only weeds and bare sand. The other side of the highway still has populations of many rare scrub endemics.

"Once the scrubs are converted to subdivisions, motor home parks or citrus groves, they will not likely be scrub again. Once scrub species go extinct, they will not be created again. Florida is filling up with people, concrete, and garbage. If we are to have them, scrub preserves must be established now."

In the end, Christman's meticulous survey of the Lake Wales Ridge scrub combined aerial photography and thousands of hours of exploration to identify more than 250 undisturbed Ridge scrub and sandhill parcels. He also documented the status of thirty-five rare scrub plants.

While the vast amount of data he collected was rowed and columned into appendixes of the report, his narrative of the otherwise data-packed 250-pager evoked the melancholy of Rachel Carson's *Sense of Wonder*, which suggested that "it is not half so important to *know* as to *feel*."

Christman's report became a rallying cry for the special gathering at Archbold in November 1989. There, scientists and conservationists from Archbold, The Nature Conservancy (TNC), and various federal, state, and local agencies itemized every single remnant tract of scrub in Christman's report and prioritized them for purchase—a dream that would lead to the development of the Lake Wales Ridge National Wildlife Refuge.

An ambitious conservation agenda put together after the meeting helped spur the Florida legislature to approve Preservation 2000, a program of the state legislature to preserve natural lands, and the sale of $3 billion in bonds for habitat protection. TNC, Archbold, and its partners then went to work trying to protect nearly two dozen high-priority tracts that could be connected by the smaller stepping-stones of scrub that had been purchased earlier.

But even as conservationists were buying up land to protect it, developers were competing to buy it up. By the early 1990s, thousands of acres of scrub along Lake Placid had been platted for development. If the land were fully developed, the folks at nearby Archbold would soon have the dubious pleasure of watching kids chase fly balls down the driveway to the station.

The developer planned to tame the wild Lake Placid scrub in two stages—first the northern three-thousand-acre parcel, then the southern three thousand acres, in a development named Placid Lakes. By the early 1990s, houses were already being built in the northern half. The southern three thousand acres comprised the last sizable patch of scrub along the southern and western shores, and it was pretty much all that stood between the beautiful wilderness of Archbold and the juggernaut of development speeding toward it. The parcel was mostly scrub and was home to fifty families of scrub-jays and more than a dozen species of endangered plants.

John W. Fitzpatrick was the director of Archbold at the time. Twenty-five years earlier, Fitzpatrick, as a biology student at Harvard University, had worked as an intern at Archbold with Glen E. Woolfenden, a professor at the University of South Florida and head of the Archbold ornithology lab. Fitzpatrick later went on to get his PhD in ornithology at Princeton University, and in 1984 he and Woolfenden published the definitive scientific work on the species, "The Florida Scrub Jay: Demography of a Cooperative-Breeding Bird." In 1988, Fitzpatrick became executive director of Archbold, a post he held until shortly before joining Cornell University as director of the Cornell Laboratory of Ornithology in 1995.

Fitzpatrick frequently spoke publicly about the importance of saving the tract. He visited the scrub, spoke with reporters, and offered photo opportunities of scrub-jays feeding from his hand or sitting on his head. He organized meetings with local, state, and federal officials.

As he told *St. Petersburg Times* reporter Barry Millman, building up the station's research was critical, but so was "managing and preserving our unique natural areas." While not exactly a firebrand, Fitzpatrick wasn't afraid to speak his mind. Shortly after his arrival, he put the southern portion of the Placid Lakes scrub high on the agenda.

In 1993 the Florida Department of Environmental Protection bought the land from the owner, Angus Tobler, for $6.8 million, mostly with money from Preservation 2000. Archbold contributed $250,000 and agreed to manage the 3,151 acres with help from the Florida Game and Fresh Water Fish Commission.

"Preserving the Placid Lakes site was one of the proudest moments of my conservation career," Fitzpatrick told me. "I bought two bottles of Dom Pérignon when that deal closed. I invited the TNC guys who worked so hard on it. One of the great moments for Florida scrub and Archbold. Effectively doubling the total size of the preserve. And having gone through so many close calls when it was going to become something else."

The Lake Placid tract was the first in what would become almost two dozen tracts scattered along seventy-five miles of threatened Lake Wales Ridge habitat. For Archbold, the Lake Placid scrub was a critical piece of the puzzle that now connected the station to a fifty-three-thousand-acre contiguous network of state- and federally protected lands.

12

SCRUB-JAYS IN THE NEIGHBORHOOD

Basic research—studies meant to gather general information about a topic rather than prove or disprove a point—has gotten a bad name. These studies are sometimes criticized as wasteful or are parodied for their seeming aimlessness or randomness, as if the first item on every scientist's research agenda is to waste their time on another's dime.

But it would be hard to formulate a hypothesis about anything without first knowing the territory. It's hard to imagine that the first ocean adventurers set out to achieve anything beyond just getting somewhere else. That's a lot of what scientists in basic research strive to do—just to get somewhere else in understanding. And as they arrive at new places of insight and map the territories they study, they can formulate and test specific ideas or hypotheses or otherwise engage in applied, as opposed to basic, research.

Although disheartened by the development on the northern three thousand acres along Lake Placid, scientist Reed Bowman of the Archbold Biological Station realized the birds there presented a golden opportunity to learn about scrub-jays living in urbanized landscapes. That was obviously an important thing to know, since urbanization was among the biggest threats to long-term survival of the species just about everywhere in the state. In fact, at the time nearly one-quarter of all Florida scrub-jays statewide lived in or near urbanized areas, and that percentage was rapidly growing.

As houses were built, Placid Lakes Estates presented a gradient of development that would give the study "high resolution." Because the density of houses would change over time, Bowman could create a time-lapse view of the scrub-jay in the developing neighborhoods of Placid Lakes Estates. In some ways, it was a dream lab for researchers—if a nightmare for the scrub-jay.

When Bowman began the study in the early 1990s, the housing gradient along Lake Placid ranged from about 180 houses per one hundred acres in the north to fewer than 10 houses per one hundred acres in the south. The six-square-mile area was home to more than 140 scrub-jay families, although Bowman's study plots included fewer than 100 of the families. He divided the study area in two—the north, with its high density of houses and 55 scrub-jay families, and the less dense south, with its 37 scrub-jay families. Then he devoted the next few years to collecting every bit of data he could.

He mapped the location of every house and scrub-jay territory within the study area. He also banded all the birds, which wasn't necessarily complicated, since many urbanized birds had become tame and readily touched down for peanuts. Once he had banded all the birds, he counted them each October, January, April, and July, enabling himself to learn about their survival and movements.

Once a year, he did a much wider census not only in the study area but in all the scrub within flying distance of Placid Lakes Estates. This included the Archbold lands, the newly established Lake Placid Scrub Tract, and outlying patches. The census included where birds were spotted in relation to their territory (if they had one), where the birds otherwise lived, and whether they were helpers, parents, or loners. In thousands of hours spent in the scrub over the next three years, Bowman and his team found every nest, recorded the number of eggs laid, the number of eggs hatched, and the number of young that successfully graduated from the nest. He was nothing if not thorough—some might say obsessed.

After Bowman amassed and organized thousands of data points, coffee became his new best friend. He sat down and analyzed the enormous number of data points he'd collected even as his team collected more. To a layperson, the terminology would have seemed incomprehensible ("Multiple regressions of bird variables on landscape characteristics and chi-square, $df = 1$, $p < 0.05$"). But to Bowman and other ornithologists, the data was the clearest evidence yet of the impacts of urbanization on the Florida scrub-jay.

Bowman found that the fewer the houses in an area, the more common was the scrub-jay—undoubtedly because less density meant more scrub. But the easily predicted outcomes of the study pretty much ended there.

As a start, the scrub-jays at Placid Lakes Estates had smaller territories than did those in the less-developed areas. And it wasn't just that there was less land to divide among the resident scrub-jay population. The peanuts and other handouts from people in the development meant the birds needed less land to forage to come up with a square meal. The impact of this cornucopia of food had other ripple effects. It changed the birds' breeding patterns. With more food readily available, pairs bred earlier, laid larger clutches, and attempted more second clutches than did the birds in natural scrub.

Yet the percentage of juveniles that successfully grew up to join the resident population was 50 percent lower than the number of juveniles raised in the natural scrub. "Nestlings need insects for food to fuel their rapid growth rates, but residential use of chemicals had reduced insect populations, so jays were feeding their nestlings peanuts, seed, and bread—real junk food," according to Bowman. Residents hoping to help the scrub-jays by feeding them were harming them. Then again, if feeding the birds meant residents became more supportive of scrub-jay conservation efforts, that could offset some of the downsides of attracting the birds to their backyards.

But of all that Bowman learned from his study, one finding stood out in its wider implications for scrub-jays on the Lake Wales Ridge and throughout their range. Suburban areas such as Placid Lakes

Estates were magnets for scrub-jays from other semi-suburbanized landscapes. When birds were raised in areas of degraded scrub with houses, they seemed to no longer recognize the healthiest areas to live in. When these birds dispersed from their less-than-ideal habitats to move to new areas, they flew right past Archbold and some of the best remaining scrub in all of the Lake Wales Ridge. They headed to places like Placid Lakes Estates.

Bowman explained: "We discovered that the only birds that were actually the immigrants into our suburban areas were from other suburban areas. And so, these were birds that were born, hatched, and raised in suburban areas, and they were fine with dispersing and settling in other suburban areas. Now, they're fine with that because that's where they were raised. But these are sink habitats, where you can't be successful enough to replace the number of birds that die. That's the classic definition of what we call an ecological trap. There are no cues that tell the birds they can't be successful there. But they settle there anyway. Now, those birds could have immigrated from a suburban area and settled in a natural area, but they didn't. And so their long-term chances of success are greatly reduced."

Anything with the name "ecological trap," whether you knew the meaning of it or not, spelled big trouble for scrub-jays, especially as more of the remaining birds fell under the spell of urban influences. What was the point of restoring areas of degraded scrub if the birds began to favor degraded and urbanized areas of scrub? As Bowman asked, "What if urban environments siphon scrub-jays from overgrown or less-than-optimal scrub, which could be restored, into urban areas, where they would eventually perish?"

Historically, scrub-jays lived in large populations surrounded by smaller ones. In the natural ebb and flow of what biologists call "source-sink dynamics," Bowman said, "small populations are always a lot more vulnerable to just sort of winking out. But in the long term, they're maintained because birds are constantly immigrating from the large population. So those small populations can go through a sort of cycle of extinction and recolonization. They may lose all their birds one year, but then a couple of years later new birds are settled there and the population is back again. Those smaller populations persist over time."

The pattern of immigration Bowman found emerging at Placid Lakes Estates was different. The surrounding small populations (known as sinks) that once fed genetic diversity in the larger populations (known as sources) were winking out and not coming back. "Once a sink goes extinct, it's going to stay extinct because the birds don't leave the source and settle in the sink. That's not sustainable."

Bowman had hard numbers to back up his conclusions. During the first four years of his study, 59 jays immigrated to and bred in the Placid Lakes Estates population. The rate of immigration was high enough to offset the high death rate from snakes and other suburban perils. Yet the suburban population still declined by 10 percent. Bowman continued the study for nearly twenty years, until 2010. "Over the twenty years I've studied Placid Lakes Estates, the population declined from 120 groups down to 7 or 8 groups, despite in-migration," Bowman said.

This finding could have big implications for the wild birds at Archbold and elsewhere. "If all the populations surrounding Archbold are declining, does that mean the number of immigrants that come into the Archbold population every year is also declining because, presumably, they're coming from those same populations?"

He answered his own question: "It turned out that was true. What was declining at Archbold wasn't our overall population but the number of new immigrants that came into our population every year." And that, he said, will eventually spell trouble for the Archbold population. Big populations depend on in-migrants from smaller populations to supply genetic diversity.

The decline of migrants into Archbold hadn't caused the Archbold population to decline yet. But it was still undermining what on the surface was a "stable" population. With fewer outsiders coming into Archbold, the birds there began to inbreed. They produced enough birds to maintain the population. But inbreeding was slowly leading to genetic deterioration. This lack of genetic diversity would eventually lead to unfit birds, and unfit birds inevitably lead to a population decline. "It might take a decade, or it might take a century," Bowman said. But sooner or later it seemed destined to happen.

Contrary to conventional thinking, small populations aren't expendable. Saving small populations of scrub-jays may be the only way to save larger ones. Bowman said, "We tend to have this paradigm in our head that we have to save the largest populations. But this demonstrates that small groups have a very important role in the overall functioning of populations. Because they provide immigrants, they maximize genetic diversity in the large populations. Large populations, completely isolated from smaller ones, suffer genetically."

"I believe that the extinction of these suburban populations is inevitable," Bowman concluded in his study. "It is unlikely that the deleterious consequences of jays living adjacent to human development can be mitigated in existing residential developments. Roads are the first infrastructure associated with new suburban developments, and these alone, through disruption of fire regimes and increasing jay vulnerability to roadside mortality, may be sufficient to depress recruitment of young jays. Add even a very few houses to the development and the rate of jay population growth plummets."

But Bowman did hold out some hope that better design of new housing developments could leave more room for scrub-jays. At the time, he envisioned that contiguous reserves within suburban landscapes would encourage the jays to nest far enough away from houses to avoid their perils. It turned out to be wishful thinking. The only contiguous "habitat" at Placid Lakes Estates would be the meandering interconnected greens and fairways at its golf course.

More than twenty years after the start of his study, Bowman was surprised by the number of scrub-jays still surviving at the housing development. "It's interesting that these birds haven't gone extinct, and I'm warning you I'm delving into the realm of speculation here. We've been doing a lot of work on personality in jays here at Archbold to try to understand the behavioral differences that characterize individuals and learn whether there are groups of individuals that all share those characteristics. Individual scrub-jays have personalities just as in any other animal, especially humans. And it's probably pretty likely—I wouldn't want to speculate which personality type—but it's quite possible there are some personality types that make them a little bit more resilient to things like suburban development.

"So one of the questions is, has there been extreme selection during these population declines for specific behaviors? Does that mean that the only birds that are left are all personality types that happen to do well in the suburbs? If that's the case over the long term, we could see growth again in those populations, but now of a slightly different type of scrub-jay. The only survivors of suburbanization may be the ones that actually could deal with it. It's amazing that after decades of development, so many still survive at all at Placid Lakes Estates."

The question remains whether that, in the long run, is good news or bad. Not all evolutionary "choices" guarantee a bright future for a species. If scrub-jays are in fact adjusting to living in neighborhoods—and that's still a big if—there's no guarantee this is a path to their long-term survival. Taking the "friendly neighbor" route may just be a shortcut to their eventual demise.

13

THE GOULD ROAD SCRUB

Reed Bowman, John Fitzpatrick, and I left the Archbold Biological Station in midafternoon, drove four miles through a corridor of citrus groves along US 27, and turned left onto a small byway called Gould Road. Soon we arrived at what Fitzpatrick called "the holy grail of ancient scrub."

"It's fun to get back up here. It's been so many years. This is just incredible. It's beautiful," he said as Bowman parked the truck at the crest of a hill. For the past forty-three years, Fitzpatrick had spent part of every spring back at Archbold, where he joined Bowman, director of the Avian Ecology Program, in the mapping study; 2018 marked the fiftieth continuous year.

We walked through the sand up a rise until we reached the ancient shoreline and then stood silently with a breeze in our faces, staring down into a blue-green sea of grapefruit trees below the ridge. The sea-surface of leaves shifted in the wind. It was 2016, but we were standing in the Pleistocene epoch, perhaps two million years ago, when ocean waves lapped the ridge. Sea level has since fallen hundreds of feet, and the shoreline is now fifty miles east.

The scrub was beautiful on that April afternoon. Pale purplish-pink flowers, clustered on high branches of sprawling bushes, festooned the sandy ridge. As we walked through the big rosemary bushes, we noticed a butterfly feeding on a ground-hugging plant covered with lavender blossoms that were held aloft amid the plant's tiny dimpled, elliptical blue-green leaves. Bowman said it was a member of the *Calamintha* genus, perhaps Ashe's savory—*Calamintha ashei*—a rare Florida endemic that the state lists as endangered. It grows in only a few places in Highlands County and neighboring Polk County, with an orphan population up in Ocala National Forest.

"This is one of my favorite spots," Bowman said as he pointed into the groves spanning the distant horizon. "To me, what I imagine here is the ridge when that was the ocean." He and Fitzpatrick debated about altitude. Neither was quite sure. They settled on "around two hundred feet."

I mentioned how odd a contrast it seemed to have the world of industrial groves gnawing at the foot of wild and ancient scrub. "You see it everywhere on the Lake Wales Ridge," Bowman said. "Housing developments, golf courses, backing right up to rare habitat."

Fitzpatrick said that if I really wanted to see a contrast, I should stop off at the massive junkyard just down the road from there. "It's built right in rare scrub," he said. "Go see that." He lamented all that had been lost, including 1,600 nearby acres of rare scrub filled with scrub-jays that Consolidated-Tomoka Land Company had cleared for a citrus grove in the late 1980s.

Today, this tiny ridgetop where we stood was about all that remained of an ancient band of scrub that, until even recent times, had stretched for several miles along US 27. Having been untouched by the sea for so long, it was among the oldest landforms in Florida. The unique community of life that evolved there is the rarest of the rare. We stood in the sum total—the 158 acres—of what was left.

"This spot is the crown jewel of the Lake Wales scrub," Fitzpatrick said. "This is where I invented the phrase, when they would say, you know, 'Well, you can't save it all.' And I would say, 'I'm sorry, there's no talk about saving it all anymore. The pizza is gone. We're talking about the crumbs left on the plate.' And this was one of the most precious crumbs."

In his 1988 report, Steve Christman wrote that the scrub in the area was "being destroyed so fast that it is impossible to keep up with the changes." In the early 1980s, he had seen hundreds of acres of white-sand scrub and rosemary balds abounding in scrub-jays all along this stretch of US 27. It included endangered flowering plants of the carrot family known as wedge-leaved button-snakeroot (*Eryngium cuneifolium*) as well as scrub plum (*Prunus geniculata*), pygmy fringe tree (*Chionanthus pygmaeus*), Highlands scrub St. John's wort (*Hypericum cumulicola*), a rare wireweed (*Polygonella basiramia*), and at least ten other rare plant species. The pygmy fringe tree and the snakeroot can still be found at the Gould Road tract, a parcel that is now part of the network of scattered scrub patches along the Lake Wales Ridge.

The last time the Lake Wales Ridge was inundated was about two million years ago, in the late Pliocene or early Pleistocene, when sand piled up in the very center of the Florida peninsula. "This part of Florida has been here for as long as there's been a Florida," Bowman said.

We walked a short way farther along the ridge. Topiary rosemary bushes stood in clumps taller than we were. Scattered among them were palmettos and other ground-hugging herbaceous plants. Bowman placed his foot near one of the tough, alligator-hide-textured "stems" of a scrub palmetto snaking by, half-buried in the sand.

"How old do you think this is?" he asked as Fitzpatrick walked over to take a look. "Probably about five thousand years old," he said. "People used to think the bristlecone pines in California were oldest. It may be what you're looking at right now. When researchers studied it, they found practically no growth rate. That's how slowly it grows." He said the stems could grow hundreds of yards across a scrub. To do that at "no growth rate" would mean they were very old for sure. It would take thousands of years.

I asked Bowman about the prospects for the scrub-jay. "It's complicated," he said. "Certainly, from my perspective of the jay's natural history, things I believe about the jay's conservation have changed. Quite a bit. I definitely have become more cynical and jaded about range-wide conservation."

"It's going extinct," Fitzpatrick added without hesitation. "This bird is on its way to extinction. It's down to about four hundred scrub-jay families on the whole Ridge."

It's far from alone in its predicament. In addition to the scrub-jay and rare plants, Gould Road is home to several rare scrub insects. This remnant tract of ancient coastline is home to one of only three known locations for the rare blue Calamintha bee (*Osmia calaminthae*). With iridescent blue wings and head, the bee relies on blue *Calamintha* as a source of food—and the plant depends on it for pollination. As we walked, Bowman pointed out the tiny leaves and stalks of a *Dicerandra* mint. "It's a fall bloomer, so it's pretty dormant right now."

We wandered for a while around the Pleistocene landscape. At one point, I looked up to see Fitzpatrick and Bowman standing among some palmettos on the high point farther along the ridge, both with hands in their pockets, gazing over the ridge. As I rejoined them, Fitzpatrick said, "Let me show you something," and led us downslope. "So this is one of the places where a bulldozer was here," he said, irritated by the partial destruction of the bald more than twenty years earlier. He said it had barely escaped at all.

Consolidated-Tomoka owned the ridge at Gould Road at the time. In his 1988 report, Steve Christman identified the scrub as one of the highest priorities for protection. Fearing that the surrounding groves would soon expand into the scrub, Fitzpatrick had arranged to meet with a company official.

"I took him out to Gould Road and showed him the scrub-jays. We were looking at the maps especially emphasizing this spot, and he said, 'Well, look, if we—we're planning on bulldozing it to plant citrus—if we wanted to save a little museum piece of it, you know, that's a heartbreaking thing to be hearing, right?'

"So I told him, 'The very most important spot is the northernmost edge of it, especially as you get up to that very high-rise there. It's got a high density of endangered plants; it's aesthetically majestic; just everything about it is beautiful.'

"Well, one Friday afternoon, it seemed like only a month after I had that conversation with him, Mark Deyrup came in breathlessly to my office and said, 'They're bulldozing the Gould Road scrub!' I just, I couldn't believe it. So I jumped in the car and grabbed my monstrous home Panasonic video camera and came out here and sure enough, the guy was bulldozing everything. He was on the tractor, and he was mowing it down. And he was starting at the exact spot I had said was the most important."

Fitzpatrick paused and pointed to the spot. "It was right in there.

"So I parked, and I turned the camera on. There were scrub-jays visible. I zoomed in on the scrub-jay, and I zoomed up to the guy on the tractor, and he's coming down toward me. He saw me filming him, and he chugged, chugged, chugged the tractor up to the hilltop, turned the tractor off, got in his car, and left.

"I got back in the office, and I called the head of the company. This was about two thirty in the afternoon, and his secretary said, 'He's in a meeting,' and I said, 'Please get him out of that meeting! And tell him he will agree to this—this is important enough to get out—and that I'm not going to get off this phone until I talk to him.' I was really, really ticked. So he got on the phone. I told him where I was, and I said—I told him what was happening here—and anyway, long story short, they never put the tractor out here again, and eventually it became part of the state preserve system."

As we walked past the bulldozer's old clearing, climbed into the truck, and headed back toward Archbold, Fitzpatrick continued shaking his head about the incident. He reminded me that on my way out of town, I "really ought to" stop and see one of the stranger sights on the Ridge, a big auto junkyard he'd spoken about earlier. I didn't understand exactly how the suggestion fit into my visit to one of the rarest habitats in North America until Fitzpatrick explained that the junkyard was right in the middle of it. "That's right. Sitting at the tip of the Lake Wales Ridge, right in the scrub where scrub-jays used to be," he said, with more amusement than anger.

Ole South Auto Salvage was right on US 27, less than a mile from Gould Road, so I drove south to see what Fitzpatrick was talking about. Although not obvious from the road, the yard sits on a sandy plateau. As with the Gould Road tract I had just visited, Ole South

Salvage's real estate hadn't been inundated by high sea level for a million years or two. Although the jay is long gone, some rare endemic plants can still be found among the forty-five acres of neatly arranged stripped automobile chassis.

The place evokes something of the macabre cemetery. Ole South Salvage is monument, museum, and memorial rolled up into one—right in the middle of the oldest ecosystem in Florida and one of the rarest in the world. A grid of alley-like roads run up and down the rows of metal carrion in various stages of decay, and some cars had been stripped to the bones. Many of the vehicles apparently died natural deaths—some had more than two hundred thousand miles on the odometers—while the fate of others, with their twisted and smashed front ends and crushed or collapsed passenger compartments, didn't leave much doubt about the fate of their occupants.

The mechanical boneyard of Ole South Salvage is a testament to a disposable world and the diversity of highway beasts that have died or models that are now extinct, most of them post-2000. Their rusting hulks are stripped of engines, wheels, rear- and side-view mirrors, and just about every other detachable, which now neatly hang or line shelves in the thirteen-thousand-square-foot warehouse for customers to browse in person or online by make and model, or age and species.

The junkyard is abutted immediately to the north and west by low white-sand hills covered in brush and a scattering of low-growing oaks, all reminiscent of the jay-filled scrub that once covered most of the southern part of the Ridge along US 27. Hundreds of acres of somewhat degraded scrub next to the yard are still up for grabs. How it will end up is anybody's guess. A scrub-jay would be wise not to bet on it. Now and then, a scrub-jay passes through the scrub adjacent to the yard, but scrub-jays haven't lived there for decades.

Aware of the ecological sensitivity of the area, the Ole South owners, the Thompson family, pride themselves on their efforts to recycle automobiles and protect the environment. When a dead car is checked, all the bodily fluids are removed, good tires sold, and bad ones recycled. Once a vehicle has sat in the yard until its bones have been picked clean, the chassis is crushed and sent to a metal recycling plant. According to the *Lake Placid News-Journal*, Ole South "destroys the preconception of what an auto salvage yard is. Ole South Auto Salvage is not a junkyard; in fact, it is a salvage yard that has earned the certification of 'Green Yard' for its efforts in recycling and protection of the environment."

The salvage yard currently covers about five acres more than it did in 2013, thanks to a rezoning Mr. Thompson obtained through the Highlands County zoning commission and local planning agency in the county seat, Sebring, in 2013. He asked and, after rather extensive paperwork, the zoning was changed from agricultural to industrial so he could expand his family operation. It wasn't much of an expansion, and the areas it expanded into had long before stopped functioning as scrub.

The owner of the salvage yard is Paul "Butch" Thompson. In 1968, he bought 80 acres of what he said was "scrub and lots of pine with some rosemary." He built a gas station and wreckage service at the intersection of Gould Rd. West and Highway 27 and lived nearby in a mobile home before building a house. Then, in the late 1970s, he got the 40 acres rezoned for the auto salvage yard and planted the rest in oranges. "There were colonies of the birds around our yard until the late 1970s," he said, referring to the scrub-jays. "You'd see them everywhere. When that scrub on the east side of 27 and south of Gould Rd. got cleared for groves, they started to go down and we didn't see them much after that."

I asked Terry Thompson, the founder's son, if he thought Ole South Auto Salvage would be expanding in the future and how the scrub-jay might affect it. Speaking with a clear affection for the bird, he said, "We don't see scrub-jay around the yard here very much. But we do see them in the buffer, and we have some natural scrub on the other side of the yard, and we see them over there. We don't plan to expand because we're pretty well maxed out. We're in a kind of permanent status without any environmental paperwork to fill out every year. So we're settled and don't have any ongoing issues."

As I drove through the chain-link gate and back onto the highway, I couldn't figure out the reason for my visit except that Fitzpatrick told me I should see the place. Was it a study in contrasts, a study in inevitability, a seminar on prioritization, or just a weird detour from the Pleistocene to the Industrial Age? If nothing else, the visit to the Gould Road scrub and Ole South Auto Salvage symbolized possible futures, with a warning: a decade from now there will be more new land zoned for salvage yards in Florida than there will be new protected areas for the Florida scrub-jay. It all made perfect sense. And absolutely no sense at all.

PART FOUR

OCALA NATIONAL FOREST

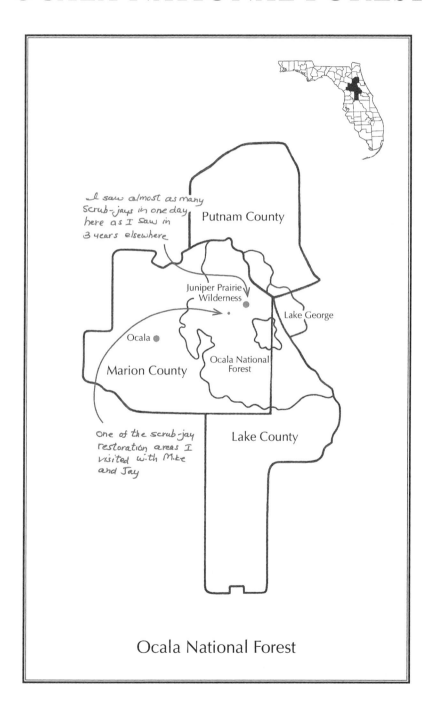

Putnam County

I saw almost as many Scrub-jays in one day here as I saw in 3 years elsewhere

Juniper Prairie Wilderness

Lake George

Ocala

Ocala National Forest

Marion County

One of the scrub-jay restoration areas I visited with Mike and Jay

Lake County

Ocala National Forest

14

BIG ENOUGH TO DREAM

Ocala National Forest, a 673-square-mile realm of protected lakes and sand-pine scrub in north central Florida, means different things to different people. To the naturalist William Bartram, who visited in 1765, the region was a "blessed land where the gods have amassed into one heap all the flowering plants, birds, fish and other wildlife of two continents in order to turn the rushing streams, the silent lake shores and the awe-abiding woodlands of this mysterious land into a true garden of Eden." Among the wonders he witnessed there were "several kinds of birds, particularly a species of jay; they are generally of an azure blue colour, have no crest or tuft of feathers on the head, nor are they so large as the great crested blue jay of Virginia, but are equally clamorous."

If, to Bartram, the Ocala region was a pristine garden, to the settlers who followed him it was a garden to be trammeled. Turpentiners slashed the trunks of pine trees and drained the sap into containers, eventually weakening the trees and leaving salvage wood behind. In other areas, settlers set fires, cleared land for pastures, and carved homesteads out of the forests.

To Theodore Roosevelt, one great belt of yet undisturbed Great Scrub between the Ocklawaha and St. Johns Rivers was a place of wild beauty to be protected for future generations. In 1908,

Roosevelt christened it the Ocala National Forest. In a 1916 report on resources within what is now Ocala National Forest, W. F. Hill, one of the first professional foresters to work there, called it a "great sand-waste area" void of agricultural value. The predominant tree, the sand pine, tended to be small by nature, took up to a half century to reach maturity in the poor soil, and eventually ended up as a natural barricade of gnarled trunks and branches. At the time, this pulp had little value for making paper, and the soil was too poor for growing crops.

By the 1940s, a new manufacturing process was invented to turn low-quality pulp into strong paper, and by the 1950s, tens of thousands of forested acres at Ocala had become gold to logging companies and pulp mills. As for the land being "protected," that was a relative word. Although the land was logged, it was highly regulated.

To lawbreakers, Ocala meant something else. The forest's isolation provided a hideaway for meth-makers and pot growers, whose illicit operations are almost invisible within the labyrinth of overgrown logging roads. The proximity to several state prisons and a federal penitentiary has added to the mystique of the "dark forest," as escaped felons sometimes hightail it into the brush, where even police dogs fear to tread. In 1994, two men recently released from prison murdered a college student and her younger brother, who were setting up their tent at a popular campground.

To US Navy pilots, the bombing range in the southern part of Ocala National Forest is a place to hone their skills. Exploding five-hundred-pound bombs and missiles sometimes hit close enough to campgrounds to startle hikers and campers, for whom the forest was supposedly a place to escape the din of Orlando and Jacksonville. The explosions often scare wild animals onto roads outside the park's boundaries and into nearby West Volusia County, prompting residents to bring their pets indoors and batten down the lids of garbage cans as wild animals seek food and refuge in residential areas.

For all these various users—naturalists, navy pilots, loggers, hikers, fishermen, and hunters—Ocala National Forest means something entirely different to the scrub-jay: it is among the bird's last best chances for surviving extinction in the mass extinction of the Anthropocene epoch.

Within the national forest—it's almost half the size of Rhode Island—lie about 280,000 acres of scrub. This is more than ten times larger than the total protected areas in all of the Lake Wales Ridge—including all the lands of the Archbold Biological Station. It far exceeds the total combined acreage of all the scrub elsewhere in Florida.

But before getting excited about how much scrub-jay habitat remains at Ocala, consider that even before Europeans arrived, the amount of Florida scrub in existence was naturally minuscule. Florida scrub, even at the height of its presettlement glory, constituted one of Earth's tinier ecosystems. By way of comparison, the Amazon rain forest ecosystem, at 1.5 billion acres, is forty times larger than the whole state of Florida. Australia's Great Barrier Reef ecosystem covers an area larger than the United Kingdom. The meager 10 percent of quality Florida scrub that has survived—Ocala is part of that—could fit into an area the size of Lake Okeechobee.

It's no surprise that Ocala is home to the largest remaining population of scrub-jays, although scientists haven't been able to accurately count them over so large an area. Ask a biologist how many, and you'll get answers like "A lot," "More than anywhere else," "A shitload," or "Who knows?"

Craig Faulhaber, avian conservation coordinator at the Florida Fish and Wildlife Conservation Commission in Gainesville, said he wasn't sure how many and that there are "only estimates floating around." The first time I met with him, Karl Miller, Faulhaber's colleague, who has extensively studied the birds at Ocala, agreed there are no precise numbers but ventured that there may be "1,000 to 1,100 family groups"—plus or minus an indeterminate margin of error.

I asked Faulhaber if it's better to have one large group, as at Ocala, or several smaller groups in different places. He said it wasn't an either-or: "While one large population is preferable to a number of small ones because that keeps the extinction probability down, a single population by itself, even if it's large, has its own problems." Besides, he said, having birds in different places wasn't just a biological necessity but also a moral obligation: "People should be able to enjoy seeing the bird elsewhere and throughout its historical range, even if just in pockets."

Todd Mecklenborg, the head of the Florida Scrub-Jay Recovery Team, agreed that the bird needs both large groups, as at Ocala, and smaller, dispersed groups to survive. His reasoning is based on the three R's of endangered species conservation.

"Resiliency, redundancy, and representation must be part of any realistic plan to save the scrub-jay," he said. "Resiliency" refers to the size of a population; "redundancy" to multiple populations; and "representation" to overall genetic diversity.

The "big" population at Ocala gives the birds a higher chance of surviving for longer than smaller populations elsewhere have. At the same time, there is a risk of keeping all the eggs in one basket. Therefore, you need redundancy, or eggs in different baskets. Preserving birds in many locations increases the species' chances of survival.

"You can't get complacent because of places like Ocala," Mecklenborg said. "That's just one population, and you don't want to see what happened to the Florida panther happen to the birds there. Panthers used to be all over [the] Southeast; now they're down to just one population in southwest Florida. The scrub-jay, historically it was in thirty-nine counties; now it's down to just two or three with stable populations. You've shrunken down the range, and that will keep happening if something isn't done, and then the scrub-jay will be facing a dire situation like the Florida panther."

But it's not enough to have both large populations, like those at Ocala and Merritt Island National Wildlife Refuge, for resiliency and scattered ones for redundancy. Historically, scrub-jays existed in both large and dispersed groups, but they also had the third R: representation. Before Europeans enlightened the "New World," the scrub-jay population consisted of a few dozen or so mostly large groups separated by rivers, wetlands, forests, and other natural barriers. Isolated for tens of thousands of years, these groups became genetically distinct. Of course, these differences didn't make them different species, just a highly diversified lot. Some of these differences were detectable, such as subtle variations in coloration, calls, and behaviors. They persist to this day in different groups. But the most significant genetic differences were invisible variations in DNA that partly reflected different adaptations to local environments. Exactly what these adaptations might be is unknown. In theory, a coastal population could be more salt-tolerant than an inland population. Or perhaps the birds in the temperate Ocala region deal with cold weather better than those in subtropical Manatee County. While these are speculative examples, the birds almost certainly evolved differently for a reason. Only their DNA knows.

But if each genetic group carries genes that make it better adapted to some environments than to others, then individuals from the populations aren't necessarily interchangeable. The best way to increase the overall chances for the birds' long-term survival is to preserve the genetic adaptations—the representation—of every group.

It doesn't take much to wipe out a genetic group. It took only one major highway, Interstate 95, to split the genetic group at St. Sebastian River Preserve State Park near Sebastian into two virtually separate populations. Given the small group to begin with, the whole population and its genetic contributions to the species as a whole now face greater peril. Some genetically unique ones probably have already been eliminated.

A natural division of populations by rivers and other barriers is one thing. Suddenly splitting a population asunder by a major highway is different. Every road creates more edge, or exposed perimeter, around a scrub. The more edge, the more frontage a neighborhood cat or dog has to get into the scrub and screw things up. Fragmentation increases the vulnerability geometrically. Any rancher knows it takes only four miles of fencing to surround a single one-square-mile pasture. Split that pasture in half by a road, and suddenly it takes six miles of fencing, even though the amount of land is essentially the same. For bisected scrub, that would be two more miles of exposed edge along which unwelcome species can invade. Fragmentation also decreases the distance from the scrub's edge to the center, putting interior nests and territories at greater risk of predators. Not to mention that roads and development can also funnel natural predators around the built-up areas and into remaining scrub.

Despite all the fragmentation that degraded scrub-jay habitat over the past century, the general contours of the largest ancient genetic groups have, apparently, persisted. Studies have shown that while development has split single populations, the resulting subgroups remain the same genetically. On a map of scrub-jays from a century ago, the distribution of genetic groups would look like the large black globs on a Holstein cow. On a current map it would look more like a dalmatian's spots. Scientists call these fragmented genetic subgroups "metapopulations."

One by one, subgroups are winking out. The loss of each puts the entire genetic group at greater risk. Sarasota County's main genetic group is in its final throes as Oscar Scherer State Park loses its last birds. Another metapopulation is biting the dust on the outlying islands of Shamrock Park, the Manasota Scrub Preserve, and Caspersen Beach, which Jon Thaxton and I visited. All the subgroups—and the entire metapopulation they composed—are already lost in Lee County, south of Sarasota, and in Levy County, just west of Ocala. Several of Brevard County's mainland subgroups are not far behind. Each of these losses destroys a potential trump card against future environmental challenges or other problems the birds may face, such as global climate change or an avian epidemic. Diversity fuels adaptation. Without diversity, a species will eventually stall and rust on the roadside of natural selection.

Biologists first began documenting separate populations of scrub-jays more than a quarter century ago, when biologist B. M. Stith identified the metapopulations and mapped their locations. In his landmark study, he used a series of metaphors to describe the vulnerability of different groups. Stith called the biggest groups, such as those at Ocala National Forest, Merritt Island, and the Lake Wales Ridge, "mainlands." He called the medium-size groups, such as those in southern Brevard County at Jonathan Dickinson State Park and in Manatee County's Bone Valley, "midlands." Finally, he described the smallest, most isolated pockets of scrub-jays as "islands."

Although Stith's metaphoric descriptions aren't exactly scientific, they provided a way to visualize a conservation strategy. First, secure the mainlands, where the bulk of the birds live. Next, create "supply lines"—habitat corridors—so birds can move from one place to another. Finally, if the birds from an island can't make it on their own back to the midlands, then stage rescue missions—translocations—to move them.

Scientists have learned a lot more about different populations of the birds since Stith's mapping. Beginning in 2006, Aurélie Coulon, a French researcher working as a postdoctoral fellow at the Cornell Laboratory of Ornithology, started studying the scrub-jays' different genetic groupings around the state—or the fragments of their original populations. By analyzing the DNA in the blood of hundreds of scrub-jays from different areas, Coulon and her colleagues created the first genetic map of the species' distribution. This enabled conservationists for the first time to focus on maintaining the critical populations representing different genetic groups—or at least representing those still large enough to stand a chance.

More recently, biologists have reconfigured the metapopulations or genetic groups into what Todd Mecklenborg calls "focal landscapes," or places with significant clustering of splintered genetic groups. Mecklenborg, who helped turn Stith's and Coulon's findings into a different conservation approach, described focal landscapes as "areas where there's still hope we can preserve land, so the outlying groups can remain connected to the bigger groups. We're trying to put out a strategy that preserves several focal landscapes throughout the scrub-jay's historic range." While Ocala will remain one of the bulwarks of scrub-jay resiliency—that is, a large intact population—Mecklenborg believes that the focal landscape approach will advance redundancy by identifying and protecting multiple populations and at the same time promote representation, or genetic diversity.

Although few wildlife officials will use the word "triage"—they prefer "prioritize"—that's what's happening out of necessity. The highest-priority focal landscape is Ocala, followed by Merritt Island–Cape Canaveral. Then comes the third-largest concentration of scrub-jays, the Lake Wales Ridge focal group, which is split by the city of Sebring. Small groups like those scattered around Sarasota will simply disappear, although some of the birds may eventually be relocated.

Whether the birds in mainland Brevard County will gain entry to the emergency room of focal landscape conservation remains to be seen. The same goes for the genetic groups of Manatee County and Moody Branch and those at Jonathan Dickinson State Park in Martin County, about fifty miles south of Sebastian. Either they must eventually register a strong pulse signifying their potential to become sources of population, or they will remain sinks and ultimately be allowed go down the drain.

* * *

The borderline group with the greatest potential lives at Jonathan Dickinson State Park. It is home to around twenty scrub-jay families. As Paul Schmalzer told me on our earlier trip through Brevard County, it is "one of the bright spots for scrub-jays on the Atlantic coast outside of Merritt Island National Wildlife Refuge." Despite the number of birds and large areas of good-quality available scrub there, the population seems moribund, perhaps because of inbreeding. That's where Ocala National Forest comes in. The abundance of birds there could mean good tidings of DNA for their impoverished brethren in Martin County. But with two hundred miles between Jonathan Dickinson and Ocala, there's no way for the scrub-jays to get there—unless they "disperse" along I-95 with the help of wildlife managers such as Karl Miller and his Ocala colleagues, who chauffeur them south in a truck or sport utility vehicle.

Nowhere else has enough birds to make it a reliable source of donors to other sites like Dickinson. Although biologists hesitate to guess how many birds live there, when I later spoke with Karl Miller and pressed him for specifics, he said there's "probably between nine hundred and one thousand family groups. That's orders of magnitude larger than anywhere else," he said. What's more, the population has room to grow. Over the next two decades, almost fifty thousand additional acres of the park will be turned from timber plantations into scrub. Miller said that "it's conceivable the population there could increase by up to 25 percent over the next twenty years."

Unlike in most other places where scrub-jays have grown accustomed to human presence, at Ocala all the birds are still wild, which, Miller said, "makes them ideal candidates to be translocated to other important populations like Jonathan Dickinson. Wild birds seem to do better after being released."

Still, moving birds from one place to another is a last resort, like an organ transplant. To be successful, translocations should mimic the natural workings of dispersal, which happens when certain individuals, for good reasons, set out from their home territories at particular times of the year. "We're focusing on birds that are less than a year old, trying to match the timing of these translocations to when natural dispersal occurs for the species," Miller said. "We've tried to time 'assisted dispersal' to mimic what would naturally happen. That's in the early spring and early fall."

Wildlife biologists have experimented with keeping the birds in cages for various lengths of time before releasing them. When translocating scrub-jays from Ocala National Forest, Miller found that releasing the birds directly into the wild instead of from holding cages produced the same results. "Wild birds don't benefit from being held in cages. So far our approach has been pretty successful, and it's cut down on excessive grooming and other neurotic behaviors they would otherwise exhibit in a cage."

The nuances of a successful relocation are endless. In addition to improved timing and choosing correct-age birds, biologists have learned the best locations for releasing them. "We do need to know where all the resident birds live because we wouldn't want to release a group of birds right in the middle of some other group's territory and cause unnecessary stress for everybody. We're really careful to map these out. But if we plunk them down in an area where there are no other neighbors, that may not work as well as moving them to a vacant patch within hearing range of some other resident birds. I think we need to give them their own space so they have the best possible chance of being accepted in a new family."

Translocations also carry risks. Most biologists argue that, whatever those are, some groups are going to perish without a genetic transfusion. New birds need to be brought to places such as Jonathan Dickinson before it's too late. If it's not already.

15

VOICES WE SHALL
NEVER HEAR

Curious to learn more about the future of Ocala scrub-jays, in 2017 I visited the national forest, the last official stop of three years of criss-crossing the peninsula looking for hope for the Florida scrub-jay. I met Jay Garcia, the wildlife biologist at Ocala, and Michael Papa, the silviculturist, in the ranger office in Umatilla, just south of Ocala. If Garcia and Papa were a little guarded at first as we sat gathered around the large wooden conference table, it was understandable. In contrast to the US Fish and Wildlife Service, the USDA Forest Service histori-cally has been an advocate for the freewheeling exploitation of natural resources rather than a steward of them. Its ethic of "sustainable use" merits quotation marks. But that reputation has begun to change at Ocala as the Forest Service has gone from being a foreman of timber harvesting to a steward of conservation for the Florida scrub-jay. The change in philosophy has been glacially slow. Papa explained why. "It literally takes an act of Congress to change land and resource manage-ment plans," he said, referring to the documents that guide the poli-cies of national forests.

Even as late as 1985, years after the Endangered Species Act of 1973 and other legislation had helped to tilt the federal bureaucracies to-ward conservation, the management at national forests was still oddly out of sync with the trend toward more sustainable stewardship. As Steve Christman wrote at the time, "It is perhaps ironic, but true, that the publicly owned scrubs on the Ocala National Forest are among those in the most critical imminent jeopardy." Not only did the Forest Service's management plan at the time call for increased use of chemical herbicides to control noncommercial plants, it also called for a 50 percent increase in logging and replanting of trees in plantations.

But in 1987 the scrub-jay got a sliver of good news when the federal government listed the bird as a threatened species. Although on-the-ground changes would be a long time in coming to Ocala, the listing put the Forest Service on notice: the bird that had once been little more than a curiosity in the Ocala National Forest now had legal standing with the Feds.

In the decades leading up to the bird's legal breakout, forest management plans barely acknowledged its existence. Except for deer and turkeys, which were valuable to hunters, foresters had managed Ocala mostly for loggers. Fires were suppressed, and thousands of acres of scrub-jay habitat had overgrown. But as the small trees gained value for making pulp in the 1950s, large areas began to be clear-cut. By the 1960s, clear-cutting was routine.

Although almost satanic in the eyes of the conservation community at the time, clear-cutting at least mimicked some of the beneficial effects of fire that scrub-jays needed to survive. This is because once loggers had cut the trees, other teams moved in with machines to remove the stumps and mechanically prep the ground, while yet another group showed up to plant seedlings. This engineered "succession" triggered a new generation of growth that the scrub-jays could use, at least until the trees got too tall eight or ten years later. This was a long way from an ideal scrub, but it was better than a forest.

The worst of the history of clear-cutting had been written in the great national forests of the American Northwest, where massive cut-downs unleashed torrential floods and erosion that ruined magnificent rivers and uprooted watersheds. In reaction to the public outcry, by the 1970s the Forest Service had limited the size of individual clear-cuts to about 150 acres. While that might prevent the razing of miles of trees along rivers in the mountains of the Northwest, the policy had an unintended consequence for the scrub-jay. Natural scrub-jay habitat didn't consist of small parcels scattered across a landscape, like the clear-cuts at Ocala. Natural fires often cleared thousands of acres at a time and left a mosaic of ragged edges and burns of different degrees.

While the small clear-cuts at Ocala might have mimicked natural burning in some respects, it left cramped squares surrounded by more mature, predator-prone forests. This kept small groups of scrub-jays effectively caged within small clear-cuts. But as new management policies began to gain traction, this changed.

At this point in the conversation, Garcia said, "Let's go take a ride," as he scooped up a few of the maps from the conference table. He said we could see firsthand how some of the changes were impacting the scrub-jays for the better. We got in the Forest Service vehicle, left the ranger station, and headed north along Highway 19 toward the forest.

Papa said that earlier policies favoring the scrub-jay were just now beginning to have a visible impact on the scrub-jay populations. In 1999, more than a decade after the scrub-jay was listed as federally threatened, the government expanded the size of clear-cuts to 800 acres, and forest managers began maintaining a younger age-class of trees—three to twelve years old—that would favor the scrub-jay instead of timber interests.

"The 1999 forest plan called for one scrub-jay management area of 1,900 acres," Papa said. "Then, in 2006, a second, 998-acre, area was added. But the culminating moment didn't come until 2016, when 52,000 acres of Ocala National Forest was officially reassigned to be managed primarily for scrub-jay and scrub species."

We were soon approaching the 1,900-acre tract. Recently cut and burned, it was sprouting three- or four-year growth. It was a huge expanse. In the distance stood a treeless ridgeline, and between that and where we stood there was nothing but newly regenerating scrub.

We looked for scrub-jays for a few minutes. Garcia was very surprised that we didn't see any. We got back in the vehicle and continued our drive.

"Reaching specific management benchmarks for the scrub-jay in a forest the size of Ocala isn't something you do on the back of an envelope," Papa said as the sandy road turned down a short but steep hill. He likened the creation of certain acreages of various age-classes of trees that occur within specific windows of time to a three-dimensional chessboard—in this case, one almost half the size of Rhode Island. "We know what we want, but we don't always know how to get there."

Fortunately, a model for the ideal landscapes Garcia and Papa were seeking to re-create already existed at Ocala, just a few miles west of where we stood. Juniper Prairie Wilderness is a beautiful thirteen-thousand-acre expanse of marshes, palm jungles, wetlands, and prime sand-pine scrub filled with happy scrub-jays. You could be a bird there and never know of the terrible plight of the species elsewhere.

It so happened I'd visited Juniper Prairie Wilderness with Steve Christman the year before. It was everything Papa made it out to be— the kind of land that scrub-jay fairy tales are made of. For a moment, I was transported back there. Scrub-jays were fluttering everywhere as I drove the sand road to the trailhead where I would meet Christman. Some flew from the scrub and briefly raced parallel to the car as if to escort me out of their territories. Scrub-jays were visible through every window, popping in and out of the roadside bushes, a few making diving flights as they crossed the sandy road. When Christman and I walked a few hundred yards through waist-high scrub along the Juniper Prairie trail, scrub-jays were in sight just about everywhere we went. Pine trees on a distant horizon disappeared as the trail sank behind a wall of low oaks. Then we were welcomed by a soft breeze moving through a peaceful grove of tall pines, just beyond the slender trunks and into a clearing, a placid ephemeral pond. We stood and chatted for a while as the scattered pines rose against the cloud-studded light-blue sky and a gibbous day-moon hung over my right shoulder. As I listened to Papa talk about what he envisioned for the

1,900-acre tract where we stood now, I imagined this world of Juniper Prairie Wilderness all over again.

Anything akin to that world that Garcia and Papa hope to re-create in Ocala National Forest is, of course, a long way off. Then again, a fifty-two-thousand-acre scrub restoration is a long-term proposition almost by definition. If not in sight during their own lives, then perhaps it will be in the next generation or two.

Garcia suddenly stopped the vehicle at the edge of the 998-acre management area to our right, as scrub-jays appeared. We got out. A sentinel flew to the top of a sand pine twenty yards from the road. He was followed by his mate and then two helpers. After a while, one of the birds sounded an alarm call. Almost simultaneously, all four birds collapsed their wings, threw their fate to gravity, and fell headfirst into the thick scrub. We had no idea what had set it off. Perhaps the bird, as a member of the avian kingdom, was, as Henry Beston wrote, "gifted with the extension of the senses we have lost or never attained, living by voices we shall never hear." We waited for a minute and then continued our way through our own little world back to the ranger station in Umatilla.

EPILOGUE

One hundred years and four thousand miles ago, this story of the Florida scrub-jay began with a drive my paternal grandfather, Freddie, took along Brevard County's Dixie Highway in his Overland 6. Freddie's journey in the 1920s took him through scrub-covered countryside teeming with scrub-jays. My own journey began in 2016, by which time almost all the scrub and scrub-jays in Brevard County had been obliterated—as they had been throughout Florida. If the destruction of scrub-jay habitat has slowed, it's only because there are fewer unprotected places left to destroy. Which brings me to the question, Where do I end a journey when no apparent end is in sight?

I'm standing atop Sugarloaf Mountain, Lake County, at 312 feet Florida's highest peak outside the Panhandle, pretty much at the center of the state. I'm alone on a bright morning under a sky of swimming-pool blue. Rising prominently above the surrounding countryside, this dramatic dome of sand was formed about two million years ago, or about the same time the scrub-jay arrived here. Once carpeted with scrub and pine, Sugarloaf Mountain is now covered in homes, yards, and pastures. From various vantage points on the mountain, I can piece together a 360-degree view of a portion of the Florida scrub-jay's historical range. What I can't see, I envision beyond the horizon. From here, I can see more than the former lands of the Florida scrub-jay. I can see the Anthropocene.

Remarkably, I stand—we all stand—on the cusp of this new geologic epoch. It's remarkable because a geologic age, after all, comes only every few million years or so. Nothing in the history of life is more momentous than those cataclysmic events that precipitate the end of one geologic era and the beginning of the next.

In the same way we divide human history into centuries, decades, and years, scientists divide Earth's history along a time scale—although one much longer than ordinary time. Unlike the orderly recurrence of celestial events that mark the passage of months and years, the geologic time scale is based on largely unpredictable events that have shaped Earth's history.

Eras, for example, are determined by major changes to the fossil record, such as the extinction or emergence of large groups of new species. Eras are divided into periods, which are based on a single kind of rock being formed at the time. In turn, periods are divided into epochs, which may span only a few tens of millions of years. On Earth's lengthy calendar, the present human moment of existence is fixed thus: the Cenozoic era, the Quaternary period, and the Holocene epoch. Our current epoch began with the end of the Ice Age, between eleven thousand and twelve thousand years ago. We are witnessing its end and the emergence of a new—and terrifying—epoch. It is the first geologic timespan characterized, in part, by one species' decimation of tens of thousands of other life-forms. Welcome to the Anthropocene.

Millions of years from now, should a technologically advanced species peer back into our time, they will see many other signs of human impact on the planet. Among them will be geologic signatures of rising carbon dioxide levels in the atmosphere, rising temperatures, the melting of polar ice caps, and rising sea level. The conversion of much of Earth's surface to row crops and the abundance of nitrogen added from fertilizer will also be evident, as will traces of radioactivity from nuclear testing. But few signatures will be more evident in the fossil record than the mass extinction. The name our species should leave behind is not *Homo sapiens*—"rational man"—but *Ego ruina*.

In haunts like the small scrub of the Enchanted Forest Sanctuary, we are in the early stages of writing our tragic story into the fossil record. And in Manatee County, what an odd anomaly the collapse of the Mosaic Company's massive gypsum stack into the aquifer will leave upon the rock layers of our time. What ghostly footprint will Kennedy Space Center leave on the ocean floor? I recall the beach where biologist Mike Legare pointed out how a recent storm had breached the artificial dune near Launch Complex 39A, an indication that sea-level rise was eroding the shoreline. The ten-year coastal floods are becoming three-year floods, which will soon be annual inundations that reach farther and farther inland and into scrub-jay habitat. Sea-level changes will be evident in the geologic record.

From where I stand on Sugarloaf Mountain, the future of destruction that will become our past is plain to see. Brevard County, which in my grandfather Freddie's day held the greatest concentration of scrub-jays on the Atlantic coast, is now home to only a few scattered groups of castaways on islands of scrub, such as the Helen and Allan Cruickshank Sanctuary or the Malabar Scrub Sanctuary.

I turn southwest toward Oscar Scherer State Park, a hundred miles distant, and recall my visit there in 2017 with volunteer Sandy Cooper and the park manager, Tony Clements. Only a dozen birds remained at the time, and Cooper felt sure they'd be gone by the following year. But in early 2017, a lone male scrub-jay made the perilous ten-mile flight from Lemon Bay Preserve across highways and the human wilds of South Venice to Oscar Scherer, took up with a newly widowed female, and produced several fledglings over the next two years. In 2019, one of those birds mated with the offspring of a different pair in the park and produced two fledglings. This was the first new family in the park in two decades. By 2019, the population of Oscar Scherer State Park had increased to about twenty birds. While propitious, the increase was a blip, not a trend, and it paled in comparison with the numbers needed to revitalize the scrub-jay population.

As one wildlife official put it, the production of a few new birds was "not significant." As Clements expressed it—somewhat more cautiously than he'd felt in 2017—"We don't have the earmarks for a sustained population here. We have about 400 acres, and you need at least 1,200. The entire county now has fewer than fifty scrub-jays, and not all of those are on protected land. Some are on small lots surrounded by houses, and Oscar Scherer will soon be completely surrounded by development." When I recently called Sandy Cooper to ask about his reaction to the unexpected increase in the birds at Oscar Scherer, I was saddened to learn that he had passed away of cancer in 2017, a few months after the bird from Lemon Bay had arrived.

The story of the scrub-jays' decline elsewhere in Sarasota County continues, and they will be extirpated soon. I remember my drive with researcher Jon Thaxton through the county's most explosive areas of development and how we fruitlessly searched through the last emerald of the necklace of coastal scrub that once draped almost the entire coastline of the county. Among the larger of the remaining gems—tiny as it is—is Shamrock Park, where two groups of scrub-jays have been hanging out since 2012. At the time of our visit in 2016, a total of six birds lived there. By 2019, a fledgling from the year before had grown up, to make a grand total of seven adults in 2019. But there were no fledglings in 2019, and that population could soon disappear.

A few clouds are now appearing on the horizon. Beneath the gathering vapor to the southeast, I envision the wellfield and the birds at Duette Preserve. I can still hear the shouts and calls at the nearby Moody Branch Preserve as excited Audubon volunteers documented the birds that morning. Manatee County remains one of the few bright spots for the scrub-jays in southwest Florida.

A hundred miles south of Sugarloaf Mountain lies the Archbold Biological Station and the birds of Placid Lakes Estates, where the genetic challenges and continuing loss of scrub on nearby private land paint an uncertain future. But the cadre of scientists at the biological station are not giving up. John Fitzpatrick and Reed Bowman will soon move into the forty-fifth consecutive year of mapping the scrub-jay territories there, studying the birds at Placid Lakes Estates, and strategizing the best way to save them there and elsewhere.

I turn 180 degrees to the north, and twenty miles away lies Ocala National Forest, a place brimming with the last best hope for saving the scrub-jays. Given the overall scrub-jay population trajectories elsewhere and the threats the birds face, if the day comes when none but a single population remains, it will be at Ocala National Forest. Should this become the only population, the scrub-jays' chances of survival will be greatly reduced.

From Sugarloaf Mountain—from anywhere, really—most striking is what you don't see. East, north, south, or west, you won't see the Carolina parakeet, among the last known wild specimens of which were documented in Brevard County in the decade before Freddie's arrival. The last-known bird of the only parrot species native to the East Coast was killed in Okeechobee County, Florida, in 1904. Two birds survived in the Cincinnati Zoo until 1918, when the male known as Incas died. Within a year, his mate, Lady Jane, had passed away.

Of course, you won't see the passenger pigeon either, whose southernmost range overlapped with the Florida scrub-jay's. Considered to be among the most abundant birds that ever lived, passenger pigeons congregated in flocks that stretched more than two hundred miles. In the early 1800s, the ornithologist Alexander Wilson wrote of "an almost inconceivable multitude" of the birds, which he estimated numbered more than two billion. Hunting, destruction of oak forests, and other factors destroyed the species in barely fifty years.

Nor will you see the dusky seaside sparrow, which once lived in the Merritt Island marshes within earshot of the scrub-jay's sandy uplands. The last dusky, which had been taken into captivity for safekeeping, died not far away in its aviary at Walt Disney World near Orlando in the summer of 1987. Its story was remarkably like the scrub-jay's. A bird of limited distribution, its decline was caused mainly by development—in this case, the impounding of marshes for mosquito control at Kennedy Space Center. Scientists had for decades counted down the dwindling population.

I'm familiar with the story of the dusky seaside sparrow because I wrote a book about its demise, *A Shadow and a Song*. In 2016, when I first met Mike Legare at Merritt Island to talk about the scrub-jay, he picked up a copy of *A Shadow and a Song* from his desk and asked, "Why don't you just change the name of the bird and republish it? Same story. Only the lead character is different."

Birds will not be the only fatalities of the Anthropocene in Florida. From Sugarloaf Mountain, you won't see the South Florida rainbow snake, which was known only from Fisheating Creek on the west side of Lake Okeechobee, south of here. You won't see butterflies such as the Florida Zarucco skipper, the Rockland Meske's skipper, and the Keys Zarucco skipper. You'll no longer see in Florida the Bahamian swallowtails or the Nickerbean blues. Of the 120 butterflies documented in the Florida Keys, several have become extinct, and at least eighteen are in imminent peril.

As I sit down at the edge of a pasture to contemplate the view, I find in my journal a few clippings of some writings I carry. One of them is Aldo Leopold's "On a Monument to the Pigeon," which he wrote for the dedication of a monument to the passenger pigeon in 1947.

"We meet here to commemorate the death of a species," Leopold wrote. "This monument symbolizes our sorrow. We grieve because no living man will see again the onrushing phalanx of victorious birds, sweeping a path for spring across the March skies, chasing the defeated winter from all the woods and prairies. . . . There will always be pigeons in books and in museums, but these are effigies and images, dead to all hardships and to all delights. . . . They live forever by not living at all."

But there is hope that the Florida scrub-jay will be remembered by living generations into the future. For now, the bird's strongholds at Ocala, on Merritt Island, and on the Lake Wales Ridge remain. Initiatives like those at Duette in Manatee County and at Jonathan Dickinson State Park on the Atlantic coast offer hope that new and vigorous populations can be reestablished. If so, the scrub-jay will be remembered not only as a species saved but as an inspiration and model for thousands of others facing a similar plight in the newest age of extinction. The Anthropocene is here. But we have a choice in how this chapter will be written—and whether it will include the Florida scrub-jay.

NOTES

Prologue

T. rex, according to recent scientific evidence: Brusatte 2018.
Those ancestors of modern birds: Brusatte 2018.
Back then, during a time of naturally high sea level: Hine 2013.
Because forests around the globe: Larson, Brown, and Evans 2016; Field et al. 2018.
This may have allowed the postapocalypse: Larson, Brown, and Evans 2016.
Whatever their secrets of survival: Claramunt and Cracraft 2015.
Among these is a branch of the family: Swain and Martin 2014; Fernando, Peterson, and Li 2017.
Accompanying it eastward were a tortoise: Swain and Martin 2014; Menges 2018.
As the fiery asteroid defined the end of one geologic age: Turney et al. 2018.

Chapter 1. Brevard County, 1925

His Overland 6 was noteworthy: Quotation from *Vero Press* 1925.
On cool mornings, breezy afternoons, and mild winter days: Harvey 2014.
A 1916 city directory described: Quotation from Hall 1999.
The birds deftly navigated around the bases: Ashton 1992.
As one researcher wrote: Quotation from Mulvania 1931, 528.
Chapman described it as "a beautiful river: Quotation from Chapman 1890, 5.
He then cut down the tree: Mingos 1923; Wright 2018.
In 1923, an article in the New York Times: Quotation from Mingos 1923.
In 1924, the curator John Kunkel Small: Hammer 2018.
This was probably because he was so taken: Quotation from Austin et al. 1987.

Chapter 2. The Scrub Whisperer

Florida Today *reporter Jim Waymer summed up*: Quotation from Waymer 2016.

Chapter 3. Island-Hopping

During the same time, the scrub-jay had declined: Brevard County, Florida, n.d.
By the time Schmalzer and I were driving the highway: Hall et al. 2014.

I told Schmalzer I'd read a piece by Representative Mark Pafford: Pafford 2016.
"Unlike the mockingbird, the Scrub-jay can't: Quotation from Hammer 2016.
On another occasion, Hammer criticized: Quotation from Hijek 2009.
Scrub-jays "in close contact with one another: Quotation from Fitzpatrick and Bow-
man 2016.
"Immediately beyond the mouth of this creek: Quotation from Vignoles 1823.

Chapter 4. Kennedy Space Center

Regarding his goal of Mars colonization: Quotation from Anderson 2014.
Despite its naturally limited distribution: Noss et al. 2015.
Breininger said that the seven hundred or so scrub-jays: Breininger 2016.

Chapter 5. Tel 4

The coastal scrub was an awesome garden: Schmalzer, Boyle, and Swain 1999.
The higher temperatures could force scrub-jays: Melillo, Richmond, and Yohe 2014.

Chapter 6. Sarasota County

With a few hundred acres of rejuvenated scrub: Thaxton and Hingtgen 1996.

Chapter 7. Death by a Million Nicks

Even by the 1920s, the birds had begun: Cox et al. 1987.
According to an article in the Sarasota Herald-Tribune: Quotation from Babiarz 2005.
The Herald-Tribune quoted Simon: Quotation from Babiarz 2005.
Thaxton argued that: Quotation from Hutchinson 2004.

Chapter 8. The Scrub-Jays of Bone Valley

According to the Florida Department of Environmental Protection: Quotation from
Florida Department of Environmental Protection 2018.
One of the world's largest phosphate mines: Pearce 2020.
Many of them had been hatched at the wellfield: Boughton and Bowman 2011.

Chapter 10. Journey to Venus

An even rarer shrub, Garrett's ziziphus: Myers and Ewel 1990.
All told, nineteen federally listed species: Archbold Biological Station 2014.
Discovered in 1962, the endangered plant: Eisner et al. 1990.
Historically, birds from the smaller outside populations: Chen et al. 2016.
While some biologists are skeptical: Houde et al. 2011.

Chapter 11. Lake Placid

Of the three million acres of xeric uplands: Turner, Wilcove, and Swain 2006.
The pittance remaining was home to some thirty species: Turner, Wilcove, and Swain
2006.

This included the scrub-jay and the Florida mouse: Weekley, Menges, and Pickert 2008.

"What I do for a living is document the extinction of: Quotations from Renner 1988.

"Development on the Lake Wales Ridge: Quotations from Christman 1988.

The parcel was mostly scrub: Schailler 1993.

Twenty-five years earlier, Fitzpatrick: Hailman, Fitzpatrick, and Bowman 2008.

Fitzpatrick later went on to get his PhD in ornithology: Swain and Martin 2014.

In 1988, Fitzpatrick became executive director: Fitzpatrick n.d.

Archbold contributed $250,000 and agreed to manage: Schailler 1993.

For Archbold, the Lake Placid scrub: Swain and Martin 2014.

Chapter 12. Scrub-Jays in the Neighborhood

In fact, at the time nearly one-quarter of all Florida scrub-jays: Bowman 1998.

Yet the suburban population still declined: Bowman 1998.

Chapter 13. The Gould Road Scrub

It included endangered flowering plants: Christman 1988; US Fish and Wildlife Service 1990.

According to the Lake Placid News-Journal, *Ole South*: Leatherman 2015.

Chapter 14. Big Enough to Dream

Among the wonders he witnessed there: Quotation from Bartram 1791.

In a 1916 report: Hill quoted in U.S. Department of Agriculture 2015.

REFERENCES

Anderson, R. 2014. "An Interview with Elon Musk about Mars: The SpaceX CEO Details His Plan to Send One Million Humans to the Red Planet." *Atlantic,* October 3. www.theatlantic.com/science/archive/2014/10/an-interview-with-elon -musk-about-mars/541539/.

Archbold Biological Station. 2014. "About Us: What Is a Biological Station?" Accessed February 18, 2019. www.archbold-station.org/html/aboutus/about.html.

Ashton, R. E. 1992. *Rare and Endangered Biota of Florida.* Gainesville: University Press of Florida.

Austin, D. F., A. F. Cholewa, R. B. Lassiter, and B. F. Hansen. 1987. *The Florida of John Kunkel Small: His Species and Types, Collecting Localities, Bibliography, and Selected Reprinted Works.* Contributions from the New York Botanical Garden, vol. 18, 204–7. Bronx, NY: Scientific Publications Department, New York Botanical Garden.

Babiarz, L. 2005. "A Bird, a Scam, and a Lot of Uncertainty." *Sarasota (FL) Herald-Tribune,* November 13. www.heraldtribune.com/article/LK/20051113/News /605243728/SH/.

Bartram, William. 1791. *Travels through North and South Carolina, Georgia, East and West Florida [. . .].* Philadelphia: James and Johnson.

Beston, H. 1976. *The Outermost House: A Year of Life on the Great Beach of Cape Cod.* New York: Penguin.

Boughton, R. K., and R. Bowman. 2011. State-wide Assessment of Florida Scrub-Jays on Managed Areas: A Comparison of Current Populations to the Result of the 1992–93 Survey.

Bowman, R. 1998. "Population Dynamics, Demography, and Contributions to Metapopulation Dynamics by Suburban Populations of the Florida Scrub-Jay, *Aphelocoma coerulescens.*" Nongame Wildlife Program Final Report. Tallahassee: Florida Game and Fresh Water Fish Commission.

Breininger, D. 2016. Personal communication.

Brevard County, Florida. n.d. "The Environmentally Endangered Lands (EEL) Program." Accessed May 10, 2018. http://www.brevardfl.gov/EELProgram/Home.

Brusatte, S. 2018. *The Rise and Fall of the Dinosaurs: A New History of a Lost World.* New York: William Morrow.

Brusatte, S. L., J. K. O'Connor, and E. D. Jarvis. 2015. "The Origin and Diversifi-

cation of Birds." *Current Biology* 25, no. 19 (October 5): R888. http://dx.doi
.org/10.1016/j.cub.2015.08.003.

Chapman, F. M. 1890. "Notes on the Carolina Paroquet (*Conurus carolinensis*) in
Florida." *Abstract of the Proceedings of the Linnaean Society of New York, for the
Year Ending March 7, 1890.* https://archive.org/details/abstractofi14188819021inn
/page/n21.

Chen, N., E. J. Cosgrove, R. Bowman, J. W. Fitzpatrick, and A. G. Clark. 2016. "Ge-
nomic Consequences of Population Decline in the Endangered Florida Scrub-
Jay." *Current Biology* 26, no. 21 (November 7): 2974–79. https://doi.org/10.1016/j.
cub.2016.08.062.

Christman, S. P. 1988. *Endemism in Florida's Interior Sand Pine Scrub.* Nongame Wildlife Program Final Report. Tallahassee: Florida Game and Fresh Water Fish Commission.

Claramunt, S., and J. Cracraft. 2015. "A New Time Tree Reveals Earth History's Im-
print on the Evolution of Modern Birds." *Science Advances* 1, no. 11 (December 11):
e1501005. https://doi.org/10.1126/sciadv.1501005.

Cox, J. A., J. W. Hardy, G. E. Woolfenden, and H. W. Kale. 1987. *Status and Distribution of the Florida Scrub Jay.* Gainesville: Florida Ornithological Society.

Eisner, T., K. D. McCormick, M. Sakaino, M. Eisner, S. R. Smedley, D. J. Aneshans-
ley, M. Deyrup, et al. 1990. "Chemical Defense of a Rare Mint Plant." *Chemoecol-
ogy* 1, no. 1 (March): 30–37.

Fernando, S. W., A. T. Peterson, and S.-H. Li. 2017. "Reconstructing the Geographic
Origin of the New World Jays." *Neotropical Biodiversity* 3, no. 1 (March): 80–92.
https://doi.org/10.1080/23766808.2017.1296751.

Field, D. J., A. Bercovici, J. S. Berv, R. Dunn, D. E. Fastovsky, T. R. Lyson, V. Vajda,
et al. 2018. "Early Evolution of Modern Birds Structured by Global Forest Collapse at the End-Cretaceous Mass Extinction." *Current Biology* 28, no. 11 (June 4):
1825–31. https://doi.org/10.1016/j.cub.2018.04.062.

Fitzpatrick, J. W., and R. Bowman. 2016. "Mockingbird vs. Scrub-Jay—Just the Facts."
Tallahassee (FL) Democrat. January 21, 2016. https://www.tallahassee.com/story/
opinion/2016/01/21/mockingbird-vs-scrub-jay-just-facts/79126042/.

Fitzpatrick, J. W. n.d. "John Fitzpatrick." Department of Ecology and Evolutionary
Biology, Cornell University. Accessed February 18, 2019. https://ecologyandevo
lution.cornell.edu/john-weaver-fitzpatrick.

Florida Department of Environmental Protection. 2018. "Phosphate." March 16. Accessed February 15, 2019. https://floridadep.gov/water/mining-mitigation/con
tent/phosphate.

Hailman, J. P., J. W. Fitzpatrick, and R. Bowman. 2008. "Obituary: Glen E. Woolfenden, 1930–2007." *Ibis* 150, no. 2 (April): 444–45.

Hall, C. R., P. A. Schmalzer, D. R. Breininger, B. W. Duncan, J. H. Drese, D. A.
Scheidt, R. H. Lowers, et al. 2014. "Ecological Impacts of the Space Shuttle
Program at John F. Kennedy Space Center, Florida." January 2014. NASA/TM-
2014–216639. Kennedy Space Center, FL: National Aeronautics and Space Administration. https://ntrs.nasa.gov/archive/nasa/casi.ntrs.nasa.gov/20140012489.
pdf.

Hall, P. J. 1999. "Index to City, County & State Directories Containing Towns in
Indian River County, Florida—K–Z." USGenWeb Archives. http://files.usgwar
chives.net/fl/indianriver/history/dirindkz.txt.

Hammer, M. P. 2016. "State Bird Doesn't Need to Change." *Tallahassee (FL) Demo-*

crat. January 16, 2016. https://www.tallahassee.com/story/opinion/2016/01/16/state-bird-need-change/78861728/.

Hammer, R. L. 2018. *Complete Guide to Florida Wildflowers: Over 600 Wildflowers of the Sunshine State including National Parks, Forests, Preserves, and More Than 160 State Parks.* Guilford, CT: FalconGuides.

Harvey, B. 2014. *Cocoa, Florida: A History.* Charleston, SC: History Press.

Hijek, B. 2009. Gun Lobbyist Shoots down Attempts to Change State Bird. *Orlando Sun-Sentinel,* September 9, 2009.

Hine, A. C. 2013. *Geologic History of Florida: Major Events That Formed the Sunshine State.* Gainesville: University Press of Florida.

Houde, A. L. S., D. J. Fraser, P. O'Reilly, and J. A. Hutchings. 2011. "Relative Risks of Inbreeding and Outbreeding Depression in the Wild in Endangered Salmon." *Evolutionary Applications* 4, no. 5 (September): 634–47. https://doi.org/10.1111/j.1752-4571.2011.00186.x.

Hutchinson, B. 2004. "Pecking Away at Jay Plan." *Sarasota (FL) Herald-Tribune.*

Larson, D. W., C. M. Brown, and D. C. Evans. 2016. "Dental Disparity and Ecological Stability in Bird-like Dinosaurs Prior to the End-Cretaceous Mass Extinction." *Current Biology* 26, no. 10 (May 23): 1325–33. https://doi.org/10.1016/j.cub.2016.03.039.

Leatherman, K. 2015. "Ole South Auto Salvage Crushes Stereotype of Junk Yards." *Lake Placid News-Journal,* February 6, 2015. https://www.yoursun.com/sebring/newsarchives/ole-south-auto-salvage-crushes-stereotype-of-junk-yards/article_74849a79-1e85-50d8-b63a-fea86664b221.html

Leopold, A. 1987. *A Sand County Almanac, and Sketches Here and There.* New York: Oxford University Press.

Melillo, J. M., T. C. Richmond, and G. W. Yohe, eds. 2014. *Climate Change Impacts in the United States: The Third National Climate Assessment.* US Global Change Research Program. Washington, DC: US Government Printing Office. http://nca2014.globalchange.gov/. https://doi.org/10.7930/J0Z31WJ2.

Menges, E. S. 2018. Personal communication.

Mingos, H. 1923. "World's Wild Life Is Fast Vanishing." *New York Times,* October 21, 1923.

Mulvania, M. 1931. "Ecological Survey of a Florida Scrub." *Ecology* 12, no. 3 (July): 528–40. https://doi.org/10.2307/1928998.

Myers, R. L., and J. J. Ewel. 1990. *Ecosystems of Florida.* Orlando: University of Central Florida Press.

Noss, R. F., W. J. Platt, B. A. Sorrie, A. S. Weakley, D. B. Means, J. Costanza, and R. K. Peet. 2015. "How Global Biodiversity Hotspots May Go Unrecognized: Lessons from the North American Coastal Plain." *Diversity and Distributions* 21, no. 2 (February): 236–44. https://doi.org/10.1111/ddi.12278.

Pafford, M. 2016. "One More Campaign: Florida Scrub-Jay for State Bird." *Tallahassee (FL) Democrat,* January 9, 2016. https://www.tallahassee.com/story/opinion/2016/01/09/one-campaign-florida-scrub-jay-state-bird/78502184/.

Pearce, F. 2020. "Phosphate: A Critical Resource Misused and Now Running Low." https://e360.yale.edu/features/phosphate_a_critical_resource_misused_and_now_running_out.

Renner, L. 1988. "Expert Documents Scrub's Demise as Bulldozers Rush In." *Orlando (FL) Sentinel,* April 18, B1, B4. https://www.orlandosentinel.com/news/os-xpm-1988-04-18-0030280137-story.html.

Schailler, R. 1993. "State Buys 3,000 Acres of Local Endangered Land." *Tampa (FL) Tribune*, 1–2.

Schmalzer, P. A., S. R. Boyle, and H. M. Swain. 1999. "Scrub Ecosystems of Brevard County, Florida: A Regional Characterization." *Florida Scientist* 62, no. 1 (Winter): 13–47. https://www.jstor.org/stable/24320960.

Shawkey, M. D., and G. E. Hill. 2006. "Significance of a Basal Melanin Layer to Production of Non-iridescent Structural Plumage Color: Evidence from an Amelanotic Steller's Jay (*Cyanocitta stelleri*)." *Journal of Experimental Biology* 209, no. 7: 1245–50. https://doi.org/10.1242/jeb.02115.

Stith, B. M. 1999. "Metapopulation Dynamics and Landscape Ecology of the Florida Scrub-Jay, *Aphelocoma Coerulescens*." PhD diss., University of Florida, Gainesville.

Swain, H. M., and P. A. Martin. 2014. "Saving the Florida Scrub Ecosystem: Translating Science into Conservation Action." In *Conservation Catalysts: The Academy as Nature's Agent*, edited by J. N. Levitt, 63–96. Cambridge, MA: Lincoln Institute of Land Policy.

Thaxton, J. E., and T. M. Hingtgen. 1996. "Effects of Suburbanization and Habitat Fragmentation on Florida Scrub-Jay Dispersal." *Florida Field Naturalist* 24, no. 2 (May): 25–37.

Turner, W. R., D. S. Wilcove, and H. M. Swain. 2006. "State of the Scrub: Conservation Progress, Management Responsibilities, and Land Acquisition Priorities for Imperiled Species of Florida's Lake Wales Ridge." Lake Placid, FL: Archbold Biological Station. https://www.archbold-station.org/documents/publicationspdf/Turner_etal-2006-StateotScrub.pdf.

Turney, C. S. M., J. Palmer, M. A. Maslin, A. Hogg, C. J. Fogwill, J. Southon, P. Fenwick, et al. 2018. "Global Peak in Atmospheric Radiocarbon Provides a Potential Definition for the Onset of the Anthropocene Epoch in 1965." *Scientific Reports* 8, no. 1 (February): 3293. https://doi.org/10.1038/s41598-018-20970-5.

U.S. Department of Agriculture. 2015. Status and management of scrub habitat on the Ocala National Forest: Landscape Scale Assessment. National Forests in Florida, Version 1, October 2015. https://www.fs.usda.gov/nfs/11558/www/nepa/102831_FSPLT3_2604372.pdf

US Fish and Wildlife Service. 1990. "Recovery Plan for Nineteen Florida Scrub and High Pineland Plant Species." Atlanta, GA: US Fish and Wildlife Service.

Vero Press. 1925. "Sebastian News." July 23.

Vignoles, C. B. 1823. *Observations upon the Floridas*. New York: E. Bliss & E. White.

Waymer, J. 2016. "You Won't Find This Titusville Mint Anywhere Else." *Florida Today*, September 2, updated September 6. https://www.floridatoday.com/story/news/local/environment/2016/09/02/rare-mint-tucked-city-wellfied-houses/89581700/.

Weekley, C. W., E. S. Menges, and R. L. Pickert. 2008. "An Ecological Map of Florida's Lake Wales Ridge: A New Boundary Delineation and an Assessment of Post-Columbian Habitat Loss." *Florida Scientist* 71, no. 1 (Winter): 45–64. http://www.jstor.org/stable/24321468.

Wright, R. 2018. "Not Quite the Last of the Carolina Parakeet." *aba blog*. American Birding Association. February 21. http://blog.aba.org/2018/02/not-quite-the-last-of-the-carolina-parakeet.html.

INDEX

Page numbers in *italics* refer to illustrations.

Millman, Barry, 109

Milner Centre for Evolution, 4

Mixer's Road Guide and Strip Maps, 20

Moody Branch, FL, *53,* 103, 136

Moody Branch Preserve, *53,* 85–86, 87–91, 148

Mosaic Company, 78–79, 80, 82–83, 147

Musk, Elon, 40–41

National Aeronautics and Space Administration (NASA): and Brevard County population growth, 31; employees of, 20, 32, 42; and land management, 42, 43, 45, 49, 50; and Space Shuttle Program, 40. *See also* Kennedy Space Center

National Rifle Association, 34

National Wildfire Coordinating Group, 50

Nature Conservancy, The (TNC), 108, 110

New York Botanical Garden, 15

New Yorker, 96

New York Times, 16

North American Coastal Plain, 41

North Merritt Island, FL, *9,* 13. *See also* Merritt Island, FL; Merritt Island National Wildlife Refuge

Ocala, FL, 96, *127*

Ocala National Forest: animals at, 140; conservation at, 139–40, 141, 143; escaped prisoners in, 130; and Florida scrub-jays, 6, 42, 130–31, 132, 134, 135, 136, 137, 140, 141–42, 148–49, 150; illegal drug production in, 130; and land management, 140, 141, 142; location of, 106, *127,* 129; and logging, 140; Navy bombing range in, 130; origins of, 129–30; plants in, 119; size of, 129, 131; staff of, 139

Ocklawaha River, 129

O'Connor, Brendan, 87, 88–89

Okeechobee County, FL, 149

Ole South Auto Salvage, 95, 122–25

"On a Monument to the Pigeon" (Leopold), 150

Orange County, FL, 107

Oranges (McPhee), 96

Orlando, FL, 105, 130, 149

Orlando Ridge, 96

Orlando Sentinel, 106–7

Oscar Scherer State Park: and Florida

scrub-jays, 57–63, 134, 147–48; location of, *53,* 57, 64; origins of, 57; as part of scrub ecosystem, 65

Osprey, FL, 64

"Our Land!" (Atwater), 74

Pafford, Mark, 34

Palm Bay, FL, 17, 37, 38

Papa, Michael, *127,* 139, 141, 142–43

Paxtuxent Wildlife Research Center, 86

Pelican Island, FL, 14, 15, 42

Peterjohn, Bruce, 87

Placid Lakes Estates, 112, 113, 114, 115, 116, 117, 148

Plants: Brazilian peppers, 66; chaparral, 5, 6, 23, 25, 28; citrus trees, 17, 75, 96; evolution of, 4; exotic, 22; flatwoods, 77, 99, 105; hardwoods, 22, 37, 105; invasive, 22, 66; oaks, 99, 149; and phosphate, 77–78; pines, 13, 17, 25, 28, 32, 37, 99, 105, 120. *See also* Agriculture; Scrub, plants in

Polk County, FL, 77, 93, 97, 119

Preservation 2000, 108, 109

Prophet, Katherine, 87

Quest Ecology, 79, 87

Rawlings, Marjorie Kinnan, 29

Rockledge, FL, 30

Roosevelt, Theodore, 15, 129–30

Roseland, FL, 13

Royal Ontario Museum, 4

Saint Basil the Great, 85

Sand County Almanac, A (Leopold), 27

Sarasota, FL, *53,* 57, 69, 76, 84

Sarasota Audubon Society, 73

Sarasota County, FL: citrus cultivation in, 75; and conservation, 64, 71–74, 136; development in, 58, 64–65, 69–70, 71, 72, 73, 77; and Florida scrub-jays, 57, 58, 59, 64, 65, 66, 75, 136, 147–48; government of, 69–70, 71–73; and land management, 60; location of, *53;* scrub in, 59, 65–66, 68–69, 77, 148; and US Fish and Wildlife Service, 70, 71, 74–75

Sarasota Herald-Tribune, 70, 72

Scherer, Oscar, 57

Mark Jerome Walters is a professional journalist and a veterinarian who has worked in Africa, Asia, Europe, South America, and throughout North America. His books have been favorably reviewed in the *New York Times, New York Review of Books, Nature,* and numerous other scholarly and popular publications. He has a B.A. in English Literature from McGill University, a master's degree from the Columbia University School of Journalism, and a D.V.M. with an international focus from the Cummings School of Veterinary Medicine at Tufts University.